D0074727

Waste Management

Recent Titles in the

CONTEMPORARY WORLD ISSUES

Series

Transgender: A Reference Handbook
Aaron Devor and Ardel Haefele-Thomas

Eating Disorders in America: A Reference Handbook
David E. Newton

Natural Disasters: A Reference Handbook
David E. Newton

Immigration Reform: A Reference Handbook
Michael C. LeMay

Vegetarianism and Veganism: A Reference Handbook
David E. Newton

The American Congress: A Reference Handbook
Sara L. Hagedorn and Michael C. LeMay

Disability: A Reference Handbook
Michael Rembis

Gender Inequality: A Reference Handbook
David E. Newton

Media, Journalism, and "Fake News": A Reference Handbook
Amy M. Damico

Birth Control: A Reference Handbook
David E. Newton

Bullying: A Reference Handbook
Jessie Klein

Domestic Violence and Abuse: A Reference Handbook
Laura L. Finley

Torture and Enhanced Interrogation: A Reference Handbook
Christina Ann-Marie DiEdoardo

Racism in America: A Reference Handbook
Steven L. Foy

Books in the **Contemporary World Issues** series address vital issues in today's society such as genetic engineering, pollution, and biodiversity. Written by professional writers, scholars, and nonacademic experts, these books are authoritative, clearly written, up-to-date, and objective. They provide a good starting point for research by high school and college students, scholars, and general readers as well as by legislators, businesspeople, activists, and others.

Each book, carefully organized and easy to use, contains an overview of the subject, a detailed chronology, biographical sketches, facts and data and/or documents and other primary source material, a forum of authoritative perspective essays, annotated lists of print and nonprint resources, and an index.

Readers of books in the Contemporary World Issues series will find the information they need in order to have a better understanding of the social, political, environmental, and economic issues facing the world today.

Waste Management

A REFERENCE HANDBOOK

David E. Newton

An Imprint of ABC-CLIO, LLC
Santa Barbara, California • Denver, Colorado

Library of Congress Cataloging-in-Publication Data

Names: Newton, David E., author.
Title: Waste management : a reference handbook / David E. Newton.
Description: Santa Barbara, California : ABC-CLIO, [2020] | Series: Contemporary world issues | Includes bibliographical references and index.
Identifiers: LCCN 2020012939 (print) | LCCN 2020012940 (ebook) | ISBN 9781440872822 (hardcover) | ISBN 9781440872839 (ebook)
Subjects: LCSH: Refuse disposal industry. | Refuse and refuse disposal. | Integrated solid waste management.
Classification: LCC HD9975.A2 N49 2020 (print) | LCC HD9975.A2 (ebook) | DDC 338.4/762844—dc23
LC record available at https://lccn.loc.gov/2020012939
LC ebook record available at https://lccn.loc.gov/2020012940

ISBN: 978-1-4408-7282-2 (print)
978-1-4408-7283-9 (ebook)

24 23 22 21 20 1 2 3 4 5

This book is also available as an eBook.

ABC-CLIO
An Imprint of ABC-CLIO, LLC

ABC-CLIO, LLC
147 Castilian Drive
Santa Barbara, California 93117
www.abc-clio.com

This book is printed on acid-free paper ∞

Manufactured in the United States of America

Americans threw away about 34.5 million tons of plastic in 2015 (the last year for which data are available). Just under 10 percent of that waste was recycled. Another 15 percent was incinerated. The remaining 75 percent—an estimated 26.2 million tons—was deposited in dumps and landfills. Those plastic wastes will remain unchanged in the dumps and landfills for generations, hundreds of years, or even longer.

The U.S. nuclear power industry has accumulated nearly 80,000 metric tons of wastes since the first power plant was built in 1958. According to the best present estimates, the industry will continue to generate at least 2,200 tons of high-level radioactive waste per year for the foreseeable future. At the present time, all that waste is stored at seventy-five reactor sites in thirty-three states where it was eventually produced. No single national storage site where wastes can be safely held for hundreds of thousands of years has yet been developed. No such site is even on the drawing boards.

What do these two snippets have to say about the waste management problem in the United States today? As different in size and scope as the examples are, they both say essentially the same thing: Americans are producing huge amounts of household waste and too much hazardous nuclear waste. And no adequate system for dealing with either of these issues has yet been developed. Thus, in a world where climate change, threats to democratic systems, widespread poverty, serious health problems, and similar issues are at the forefront of people's minds, wastes are probably mostly a second thought.

But for those individuals and companies actually involved in the waste management industry, those problems are certainly not insignificant; in fact, they pose a daunting challenge for communities, businesses, specialized fields of waste management, and yes, even individual people. The most basic question of all is this: How are those wastes supposed to be disposed of? For many centuries, the answer to that question has been some version of "throw them out the window," "dump them in an empty field," "toss them in the nearest river," or some variation of these themes. Even today, untold numbers of open dumps remain as the main source of waste disposal in the world, and some thousands still remain in the United States. A more developed form of disposal—the sanitary landfill—has now become the disposal system of choice in the United States, with about 2,200 such sites open for business. The bottom line, however, is that landfills are no longer an adequate tool for disposing the huge amounts of solid wastes our society produces, not by a long shot.

Discussions about waste management now appear to be going in the direction of a concept known as integrated solid waste management (ISWM). Within that concept, all aspects of waste management—generation, transportation, storage, recycling, landfills, incineration, and composting—are considered parts of a single system. Not one of these steps occurs without consideration of the way it interacts with the other steps in the process. A detailed discussion of this way of thinking about wastes is the core approach of this book. In addition to learning about the history of waste management and its current status, the book also reviews some current problems in the field, such as exportation of hazardous and plastic wastes; the status and future of plastic wastes, nuclear wastes, electronic wastes, industrial wastes, agricultural wastes, and mining wastes; and environmental justice.

The book is intended to be a research tool for readers. Among the resources to be found here are a collection of essays discussing current developments in the field (chapter 3);

review of some important individuals and organizations in the field (chapter 4); selections from some important documents in the field (chapter 5); an annotated bibliography of useful books, articles, reports, and internet sources (chapter 6); and a chronology of events in the history of waste management (chapter 7).

Waste Management

BOOKBINDER

Waste is an inevitable byproduct of society. (Geoscience for America's Critical Needs. 2016)

Yes, that's almost certainly true. But, on the other hand:

> The United Nations [has] warned that poor waste management threatens human settlement globally. ("UN Warns Poor Waste Management Threatens Human Settlement Globally" 2019)

So the problem we face is that the production of wastes is unavoidable in human societies simply because people are alive and carry out a host of normal activities. And the more advanced societies become, the wastes they produce increase in quantity and complexity. Yet humans have also known for as long as civilization has existed that wastes are undesirable. They damage the environment and human health . . . and they usually smell bad! (As we shall see, no small consideration.)

Definitions

Words such as *garbage, refuse, slop, swill, detritus, rubble, flotsam, jetsam, sewage, ruins* . . . and the list goes on . . . are part

"White wings" sanitation workers sweep garbage from the streets of New York City during a garbage strike in November 1911. (Library of Congress)

of the common parlance in speaking about wastes. Most words are poorly defined from a technical standpoint but are generally *well* understood by an average person. Specialists in the field of waste disposal and management, however, use the basic terms of waste management more precisely. The term *waste*, for example, is reserved for materials or objects for which no further use is expected. By contrast, the term *refuse* refers to materials or objects for which some further use is expected, either its original use or its some new use. Even more exact terms that one might come across include *primary refuse* (materials or objects found in the place where they were originally used and discarded); *secondary refuse* (materials or objects found in some location other than that in which they were first used and discarded); and *tertiary refuse* (secondary refuse that has been moved to yet a different location).

Wastes can also be classified in various other ways. For example, they always occur in one of three states: solid, liquid, or gaseous. An example of solid waste is the garbage that everyone throws out, usually on a daily basis. The water used in the fracking process for obtaining natural gas from the earth is an example of a liquid waste. And the carbon dioxide, nitrogen oxides, and sulfur dioxide expelled from factories are examples of gaseous wastes. For this reason, discussions of waste management can be divided into three categories, one for each of the states in which wastes occur. This book, for reasons of space only, focuses almost entirely on the topic of *solid* waste management. Other books of equal length could also be devoted to other forms of waste management problems, such as water pollution and air pollution, for which many resources are available. (See, for example, McMillan 2019; Smedley 2019; Tiwary, Williams, and Colls 2019; Wakeling 2019.)

The term *solid waste* refers to, well, waste that occurs in a solid form. A very common variation of the term is *municipal solid waste* (MSW), which is produced by day-to-day activities of human life. It includes materials such as food waste, grass clippings, used clothing, bottles, all forms of household plastics, newspapers and other paper products, packaging materials, furniture, household cleaning products, and paints. Wastes that are generally not included in the MSW category are agricultural and industrial wastes; medical, electronic, and radioactive wastes; and sewage sludge.

One of the most common methods of classifying wastes is according to the source from which they come. Examples of these categories include:

- *Electronic wastes* (also known as *E-wastes*), which comes from a host of electronic devices, such as computers, telephones, answering machines, radios, stereo equipment, video cassette players and recorders, compact disc players and recorders, and calculators.
- *Hazardous wastes* are forms of solid wastes that are toxic, flammable, corrosive, or reactive. Many common household products are classified as liquid wastes, such as drain cleaners, furniture polish, silver polish, and mothballs. The category extends to far more dangerous materials, such as the radioactive solid wastes produced during nuclear power generation.
- *Medical wastes,* as the name suggests, are any by-product of the diagnosis and treatment of human and animal ailments, such as blood, microbiological cultures and stocks, human or animal tissue, used bandages and dressings, and discarded gloves.
- *Sharps,* such as needles, blades, broken glassware, and microscope slides, may be either classified as medical wastes or placed into their own category as *sharps waste.*
- *Universal wastes* are products that "seem to come from everywhere," such as batteries, fluorescent lamps, products containing mercury and nonempty aerosol cans.

Finally, two fundamental terms associated with wastes are almost certainly self-explanatory. *Waste disposal* refers to the processes by which used objects and materials are generated and then disposed of. *Waste management* concerns all the procedures and mechanisms by which waste products are generated, converted, and eliminated, or otherwise handled by humans. *Waste prevention* and *waste minimization* involve all the methods that have been devised and used to reduce the amount of wastes produced. And *waste diversion* refers to all methods by which wastes can be prevented, reduced, and used for other productive purposes.

Waste Disposal and Management in History

The first sentence of this chapter says it best: wastes occur because humans live. When humans eat, they throw away unused foods. When they make tools and weapons, they throw away flakes that are by-products of the operations. When they construct buildings, they throw away bits and pieces of materials not used in the process. And the more complicated human life becomes, the wastes they produce increase in quantity and complexity. Stone Age humans produced wastes. But those wastes did not include electronic wastes or radioactive wastes.

Archaeologists have been studying the topic of wastes in prehistory for decades. Their task is made especially difficult, of course, because many of the objects and materials one might expect to find no longer exist in any form. Clothing, food, feces, and other organic materials have long since decayed and disintegrated. Researchers are left, then, with a variety of inorganic substances from which to determine whether ancient peoples produced waste (yes, they always did), what kinds of wastes they generated, what these wastes tell us about prehistoric lifestyles, how early peoples dealt with these wastes, and so forth. (See, for example, Amick 2014.)

Given these challenges, it is somewhat remarkable that researchers can now tell us so much about the role of wastes in prehistoric times. Some objects and materials occur frequently in one or another part of a settlement. Stone flakes are one example. As a person chipped away at a stone to make a knife, for example, flakes were produced and either left at the workplace (the so-called drop zone), or tossed away at some distance (the *toss zone*). Mussel shells are another example of a material that has been found, largely intact and undamaged, at many settlement sites. A third common waste consists of animal bones. Accumulations of these materials suggests the existence of dumps or landfills similar to those that exist today (Havlíček 2015; Havlíček and Martin 2017a, 2017b).

The problem is that one cannot know whether or not these sites are really dumps or landfills—that is, whether or not they were chosen to be "graveyards" for worn-out tools, waste shells, and unwanted or unneeded bones. Research makes it clear that these objects and materials were often recycled and used for new and different purposes. Animal bones have been found, for example, as structures from which houses, sacred buildings, meeting places, or other structures were constructed. Less commonly, it is possible that found objects and materials are truly refuse—that is, substances thrown away because they could not be used for any purpose. A broken pot or damaged knife might be examples of such instances (Andersson et al. 2004). Recent research has also found evidence for what may be the first polluted river in history. Researchers reported in 2016 that they had found waste materials that almost certainly came from very early mining operations in the southern region of modern-day Lebanon. They found samples of slag—the solid waste resulting from metal smelting that consisted of lead, zinc, cadmium, arsenic, mercury, and thallium—in addition to copper itself, the intended product of the process (Grattan et al. 2016).

Early Civilizations

The topic of waste management in early civilizations has been the subject of considerable archaeological research. One of the first discoveries made is perhaps somewhat self-evident: waste disposal posed a fundamentally different problem for villages as compared to the first large urban areas. In the former case, population density was low, and residents had adequate room to dispose of their wastes within short distances from their homes. As the population density increased, and cities of 10,000 or more began to appear, waste disposal posed a new problem. Simply dumping one's wastes into the backyard or the land next door was no longer possible. Waste dumps of some kind became a necessity. These dumps, furthermore, had to be built at some distance from the center of the town, at least on its outermost perimeters or further out. (Archaeologists use the term *midden* to describe any type of refuse heap.)

Archaeologists have now discovered several examples of such waste depositories in regions covered by ancient Mesopotamia (about 5000–3500 BCE). In most cases, these sites were nothing more than large holes in the ground, lacking the sophistication of layered structures (landfills) that first appeared many centuries later. As an example, one midden, called Tell Majnuna, was about seven acres in size, up to twenty feet in depth, and filled with about 6.5 million cubic feet of wastes. It contained a broad array of trash, such as pottery shards, broken tools and tablets, domestic refuse, figurines, kilns, ashes, clay models and containers, and materials and objects of other types (McMahon 2015, 2016).

Other early civilizations took a somewhat different view of wastes. Sinologists, for example, have often noted that Chinese civilization throughout the ages has taken the view that "people [should] do their best, make the most advantage of land, make the best use of articles, and make the smoothest flow of goods" (Lu 2018). As an example, it appears that the ancient Chinese used worn clothing in the papermaking industry ("From Waste to Resource: An Abstract of '2006 World Waste Survey' " 2016). One author has described this situation as follows:

> There were no wastes in ancient China, either in towns or countries, all wastes were degradable and used as fuel, forage or fertiliser for local ecosystem. Chinese people have a long ecological tradition of efficient resource use including wastes recycling and goods repairing. Wasting grain, paper, and clothes, no matter how plentiful they are, is considered immoral behavior that will be punished by God according to ancient tradition. (Rusong 2002)

One of the most common and most revealing lines of archaeological research on wastes has to do with the remains of mining and metalwork during the Bronze Age. This period marked the beginnings of human understanding of the nature of metals and the ways in which they could be worked to make useful products. These processes also have the benefit for modern researchers that they left abundant examples of long-lasting wastes that could be studied. Two sites in Western Europe have produced especially interesting results of mining and smelting operations: the Alps and Northern Italy and the Western Iberian Peninsula. Both locations contain waste sites that contain finished articles as well as scrap materials that provide an insight into characteristic features of the peoples living in these regions in about 2000 BCE (Havlíček and Martin 2017a).

Excavations in Egypt have disclosed that waste management systems developed in Mesopotamia were to be found in Egypt at about the same, or a later, time. The archaeology of Egyptian waste systems differs from that of other civilizations, however, in at least two ways. In the first place, the climate in Egypt is, and has long been, so arid that otherwise vulnerable materials have actually survived in refuse heaps. For example, excavations in the ancient city of Oxyrhynchus have revealed papyruses containing poems of the poet Sappho, fragments from the pseudoepigraphal biblical Gospel of Thomas, and bits of sheet music. One researcher has said that these discoveries have turned out to be "the closest thing we have to discovering the Library of Alexandria in a landfill" (Stott 2016). The other important difference was that ancient Egyptians (as is often the case today) used irrigation canals and the Nile River as waste disposal sites. These canals were so common throughout the nation that it was easy for residents simply to dump their waste products into the nearest waterway, often with unfortunate human health problems as a result (Lehne and Tavares 2010; Shaw 2012).

One other waste management practice, adopted perhaps for the first time in Egypt, was that of trash collection. Some evidence suggests that special attention was provided to the nobility and families of wealth in about 2100 BCE in the ancient city of Heracleopolis by having their waste materials actually picked up "at their doors" (as is the common practice today) and transported to the Nile for disposal. Families of lesser ranks, however, were still required to carry out waste transport on their own (Melosi 2005, 3).

The handling of refuse by the early Jewish civilization is not well documented. Perhaps the most commonly mentioned example of the practice is found in the Bible, where a reference to *Gehenna* appears about a dozen times. The term refers both to a geographical region and to a large refuse pit found there. The pit is said to have been a site at which nearby residents could dispose of their solid wastes. It was on fire constantly, producing an effect that many writers later compared to hell itself. That connection was further reinforced by reports that the site was also used for the ritual sacrifice of children from the area and for the disposal of bodies of individuals who were thought to be "unclean" for some reason or another (Loewen 2018; there is, however, some dispute as to the reliability of these reports. See "Was 'Gehenna' a Smoldering Garbage Dump?" 2011).

Disposing of human wastes was also a problem for early civilizations (as it has, of course, always been for human communities). In prehistoric times, many people probably went into the woods, behind a bush, or to some other private place to defecate and urinate. (Such practices have certainly not disappeared.) The World Health Organization estimates that about 4 billion people worldwide (more than half of the world's population) still do not have sanitary toilets available for their use ("2.1 Billion People Lack Safe Drinking Water at Home, More Than Twice as Many Lack Safe Sanitation" 2017). And archaeological evidence shows that many people in very early civilizations used simple holes in the ground as toilets (Tiwary and Saurabh 2018).

Within at least some early cultures, however, rudimentary toilets were apparently in wide use. In the ancient cities of Harappa and Mohenjo-Daro in the Indus Valley, for example, nearly all homes and other buildings structures had developed toilet systems as early as the third millennium BCE. The systems consisted of a place to sit, usually made out of stone, placed above a cesspool for the collection of wastes. For members of the upper class, this system was placed inside a home, with the cesspool being emptied or cleaned out by slaves on an every-few-days basis. The system may also have contained an outflow system consisting of stone pipes leading to an exterior drainage system. For the less wealthy and poor, similar systems were built outside the home, a primitive type of outhouse (Antoniou et al. 2016; this resource also contains photographs of a variety of early toilet systems). In one of the earliest examples of solid waste recycling, the human wastes collected by one of these systems was often then deposited in fields outside city walls, where they were used as fertilizer for crops. In other cases, toilet systems were actually built *within* pigsties, partly to cloak the unpleasant odor they produced and partly as a source of food for the pigs (Antoniou et al. 2016).

Ancient Greece and Rome

By the time the earliest modern civilizations had begun to appear, perhaps around 500 BCE, the fundamental technologies of waste management had been developed. Home-based systems had been developed for the collection and removal of personal wastes, such as wastewater, urine and feces, and kitchen and other food preparation wastes. Sewage systems to transport such wastes from individual homes to some general collection area, such as a nearby lake or river, were relatively well developed and in place in most cities and towns of any size. Community waste management systems were also common, if simple in design. People in cities were generally required to transport their household or other solid wastes to dumps, landfills, or other sites, most commonly at a location outside city limits. In many cultures, but varying in frequency from one culture to another, composting and recycling was also common. The ancient Greeks, then, were familiar with and put into practice many of the basic principles of waste management in use in today's world. The history of waste management from that point on, then, consists largely of the invention of new waste management technologies, such as the incineration of wastes in the 19th century; the improvement of traditional technologies, such as the development of sanitary landfills in 1912; and the development of laws, policies, and practices that dictate a variety of waste management systems.

As an early example of the evolution of waste management policy, the Council of Athens enacted a comprehensive ordinance sometime around 500 BCE dealing with the collection and disposal of trash in the city. The ordinance required the creation of what might be called the world's first municipal dump at a distance of at least two kilometers (one mile) from the city's outer boundaries to which all trash had to be delivered. Provisions were also made for the hiring of individuals to sweep the city's streets and to separate reusable materials from refuse in the garbage. The council edict also prohibited the disposal of wastes into the city's streets, apparently one of the first such laws in municipal history. One expert notes that the city dumps were also used for the disposal of unwanted babies! (Medina 2007, 19–20; Melosi 2005, 4).

Additional advances in waste management occurred during the rise and flourishing of the Roman Republic and Empire from about 500 BCE to about 500 CE. During this period, Rome itself became one of the world's first true metropolises with the population in the city itself reaching perhaps a million. The types and degree of waste management in the urban area itself (this discussion does not cover the widespread reaches of the empire) depended heavily on one's social status. Among the poorer families, waste disposal practices were as simple as in most previous civilizations. Human and household wastes were disposed of by throwing them out of an upper level of the house or by dumping them directly on the street in front of the house. In some cases, refuse was simply laid aside on the floor of the house itself. It was the residents' responsibility to collect wastes wherever possible and transport them to storage areas outside the city (Havlíček and Morcinek 2016).

Roman law attempted to deal with some of the most egregious examples of waste disposal. Dumping was normally prohibited near public fountains, with notices of such restrictions posted at those locations. Violators of these laws were fined or, in the case of slaves, physically punished. Owners of buildings were also made responsible for keeping streets in front of those structures clean and clear of refuse. The actual process of maintaining clean streets (for which building owners paid a fee) was carried out by teams of collectors who swept or shoveled up wastes, loaded them into carts, and transported to dumps outside the city limits. This system is sometimes said to be the world's first organized garbage collection system ("The Past, Present, and Future of Solid Waste Disposal" 2017).

Dump sites around Rome sometimes grew to enormous sizes. The largest known site was called Mount Testaccio, located near the Tiber River. It had (and still has) a circumference of about a half mile, with a height of about 115 feet. The refuse within the mound contained a volume of about 760,000 cubic yards, consisting almost entirely of more than 55 million damaged amphorae. The structure is still easily visible today ("Exhibition: Mount of Amphorae" 1997).

A key feature of the Roman waste management system was its sewer system. Specially constructed systems of pipes and canals had been in use for centuries before the rise of Rome, but they seldom, if ever, reached the level of complexity and efficiency as the Roman system. That system consisted of four levels: individual channels leading from separate buildings that fed into street sewers, leading into larger major sewers, and ending finally in collector sewers. Users of the sewer system paid a fee for the privilege, so this mode of refuse disposal was not available to all residents of the city. Pipes in the system had a tendency to become overloaded and clogged, requiring regular clean-outs by teams trained for that purpose. In some areas, the sewer pipes were so large (up to twelve feet in diameter) that cleanup crews could actually travel down their length in boats (Bradley 2012; see photo, p. 78).

This system transferred all types of wastes from private buildings to major waterways, such as the Tiber River itself. While an efficient system in and of themselves, the sewers did not necessarily guarantee that streets would be clear of refuse. In fact, many streets had stepping stones to be used by pedestrians to keep them from walking on streets filled with human feces, urine, and household wastes.

A technological breakthrough of substantial importance to the history of waste management, as well as that of Roman history itself, was the development of plumbing technology in about 200 BCE. Prior to that date, aqueducts, sewer systems, and other elements of the water supply and disposal system were made of stone. Not only was stone a somewhat unstable material, it was also difficult to arrange so as to produce an efficiently functioning system for moving fresh water and wastes. In a great discovery about which we know relatively little, Roman engineers discovered that pipes made of lead metal were easier to build and maintain than traditional stone systems. The water-movement system that has survived to the present day, plumbing, was, in fact, named after the Latin word for lead, *plumbum*. The presence of the toxic metal lead has long been blamed for an epidemic of mental and physical diseases in Rome, perhaps accounting for the downfall of that civilization. Considerable dispute exists as to the likelihood of that scenario, however (Newitz 2017).

Overall, waste management in Rome reached by far its highest level in human civilization, although access to that system differed significantly depending on one's social and economic level.

The Middle Ages

"The poor you will always have with you." This comment is attributed to Jesus in the New Testament. And yes, the rich will also be around. But there is a difference. Poor people have lived under horrendous environmental conditions throughout most of history. If one could go back to the days of ancient

Babylonia, for example, "normal" living conditions for the poor probably meant a simple shack with a dirt floor and grass roof, situated on a muddy plot of land surrounded by filth of all kinds. Disease was probably rampant, and life was short for everyone living there. And a visit to the poor of ancient Greece, Rome, China, or India or modern-day Eritrea or Honduras would probably not be much different. Meanwhile, the rich in every society have always found ways to live safe, clean, comfortable lives that included, for example, sophisticated toilet systems that their servants flushed for them (Antoniou et al. 2016).

That having been said, most experts agree that waste management technology declined significantly in the transition from Roman society to the Middle Ages. Living conditions for both the rich and the poor were substantially more difficult between about 500 CE and the middle of the 19th century. As one expert has written, at the end of the Roman Empire, "the taps were being turned off all over Europe; they would not be turned on again for nearly a thousand years" (Wright 1960). And, as another writer has noted, this change meant that "[s]anitation technology had entered its dark ages" (Kahn, Allen, and Jones 2007).

The invasion and destruction of Rome by barbarian tribes had two major effects with regard to waste management systems. In the first place, those episodes resulted in the destruction of many physical elements of the Roman sanitation system. All aqueducts bringing clean water to the city itself were destroyed as were many public baths and the city's extensive sewer system. Also, the marauding invaders destroyed the Roman political and economic system to the extent that the administrative structures and financial systems needed to rebuild the sanitation system had been destroyed. These effects were quite different in the western and eastern parts of the empire. In the latter case, widespread damage was much less severe, and the young Islamic civilization brought with it the intellectual, social, economic, and political structures needed to maintain advanced elements of the earlier Roman systems (Rosen 1958, chapter 3, 18–36).

Many descriptions of waste disposal conditions in various parts of Western Europe during the period are available. For example, one section from the *Zbraslav Chronicle*, a history of Bohemia in the 13th and 14th centuries describes the experiences of at least some people in the region at the time:

> When the poor where turned away from the doors of Prague houses and were not admitted for lodgings due to the thefts they had committed, at night lying on the streets and squares, they, for the nakedness of their bodies and the chill of the frost, would climb, like pigs, into dung that had been thrown from horse stables into the street. (Quoted in Havlíček, Pokorná, and Zálešák 2017, 272; see also Lofrano and Brown 2010; "The Middle Ages 'Roots'" 2019)

Almost certainly, the worst consequence of having solid wastes dumped into the streets and local waterways was the easy spread of disease among the populace of an area. A host of often fatal diseases was common during the period, including dysentery, diphtheria, gonorrhea, influenza, leprosy, cholera, malaria, measles, smallpox, typhoid fever, and perhaps worst of all, plague. Plague is an infectious disease caused by the bacterium *Yersinia pestis*. The disease is transmitted among humans most commonly when a person is bitten by a rodent flea carrying the bacterium. Three types of plague exist, the most common of which is the bubonic plague, transmitted by flea bites. A second type of plague, pneumonic plague, is transmitted through the air when one is exposed to the droplets released when an infected person coughs. The third type of plague, septicemic plague, is transmitted through flea bites or contact with an infected person.

If one imagines the living conditions of humans during the Middle Ages, the chance of contracting the disease would obviously be quite great. Given the profusion of refuse in city streets, the rat population must have been very large and the chances of being bitten very significant. Also, the cause of the disease was not known until 1894, so no treatment (other than prayer) was possible. Once a person was infected, symptoms included fever, headache, chills, weakness, and swollen lymph nodes. Black splotches appeared on the body because of broken blood vessels beneath the skin, accounting for the disease's common name of the *Black Death*. The disease progressed rapidly, and most victims died within a few days (Ziegler 2017).

Three episodes of the plague occurred. The first, called the Justinian Plague, is thought to have originated in Egypt before spreading to Constantinople in 542 CE. The disease spread through Europe over the next two centuries before it disappeared in about 750 CE. By that time, it may have killed as much as half the population of Europe.

The second, and probably the best-known epidemic, was the Great Plague, or Black Death, thought to have originated in China in 1334. It spread to Europe in about 1347 by way of trading activities. It lasted for four years and killed up to 75 million people. The third epidemic, called the Modern Plague, struck China in about 1855 and swept across that country and India over the next century, officially being declared "ended" by the World Health Organization in 1960. By that time, an estimated 12 million people had died of the disease. This brief description of the disease does not begin to express the social, cultural, political, economic, and other changes that it brought to Europe, parts of the Near East, and most of the Far East (Klick, n.d.; Routt, n.d.; Zapotoczny 2006).

So, what do the plague and other diseases have to do with waste management? The underlying problem in the diagnosis and treatment of disease, of course, was, and always has been, the *cause* of the disease. Medieval physicians had a variety of opinions on the matter. Many looked to the theories of Hippocrates, more than 1,500 years earlier, and thought that an imbalance of bodily "humours" was responsible for illness.

(Bleeding was the most common treatment developed as a result of this theory.) Another widely accepted explanation for disease had a theological basis—namely, that illness was God's punishment for evil acts. Under this theory, the most promising treatment consisted of prayer and other acts of religious atonement. One of the most common theories said that disease resulted from *miasma*, a noxious form of "bad air," sometimes referred to as "night air." The theory itself occurred in a variety of forms, one based on the notion that the realignment of the planets caused a change in Earth's atmosphere, so astrological predictions were needed to study disease. Other theories blamed rotting organic matter for an increase in miasma and a corresponding spread of disease.

Yet another theory of disease, poorly understood and not widely accepted, was based on the notion that illness was caused by "particles" or "seeds" that were transmitted from a sick person to someone who was otherwise healthy, or from an infected animal to a person, an idea that was later to become known as the Germ Theory of Disease. That theory did not become widely popular until the mid-19th century, however, and did so only as the result of new and increasingly convincing evidence about the relationship between disease and wastes ("Medicine and Health in the Middle Ages" 2019).

The Age of Sanitation

One of the most important breakthroughs in the understanding of infectious diseases occurred in Great Britain in 1842 with the release of a "General Report on the Sanitary Conditions of the Labouring Population of Great Britain." The report was written by attorney and social reformer Edwin Chadwick. Chadwick paid out of his own pocket for extensive research on the connection between human disease and the treatment of wastes in the city of London. His report contained several recommendations for changes in the latter, such as the implementation of systems for the removal of refuse from homes and streets, the provision of clean water for all homes in the city, and improved drainage systems for the town. His report provided the basis for the passage of the British Public Health Act in 1848, an act that incorporated Chadwick's recommendations, along with several other provisions for improved waste management systems in the country ("Campaign for Public Health (Summary)" 2019; "The 1848 Public Health Act" 2019). Ironically, in spite of these accomplishments, Chadwick remained a firm believer in the theory of miasma throughout this life and advocated these changes only as a means of reducing the cost of medical treatment for the poor by the national government.

The Public Health Act of 1848 was only the first of a long series of acts by which the British government expanded and improved its laws for managing wastes. Three decades later, for example, the parliament adopted the Public Health Act of 1875, which dramatically changed the focus of waste management programs, from national to local activities. The act provided funds to local communities to create and repair sewer systems, to develop waste and fresh water systems, and to develop new laws for the construction of private residences and commercial buildings to include waste management systems (Public Health Act, 1875).

Probably the most famous episode in the burgeoning field of sanitation research in Great Britain was the work of English physician John Snow. In 1854, a cholera epidemic broke out in the Broad Street area of London. Suspicious that the miasma theory of disease transmission was inadequate to explain this event, Snow decided to conduct a series of experiments to see if he could locate another cause for the epidemic. He eventually discovered that the water used by residents of the area for drinking, cooking, washing, and other everyday activities came from a well standing no more than a few feet away from a dump site for human and other wastes. Snow concluded that some agent in the waste pit migrated into the fresh water supply, carrying cholera to users of the pump. He recommended that authorities remove the handle of the pump. When they did so, the disease was very soon eliminated from the neighborhood.

The Public Health Acts of 1848 and 1875 were only modestly successful, at least partly because citizens appeared to not take them very seriously and partly because of the decaying condition of the London sanitation system. As was the custom in all urban areas, waste products were almost always dumped into sewer systems that, in turn, emptied in major waterways. In London in the mid-19th century, this situation was made only worse by a rapidly growing population and the invention of the flush toilet. The combination of these two changes greatly magnified the problem of waste disposal in the world's largest city at the time.

That situation came to a head in the summer of 1858 when unseasonably hot weather exacerbated the problem of waste smells escaping from the Thames River, The period became known as the Great Stink and was described by the renowned English scientist, Michael Faraday, in the following words:

> The smell was very bad, and common to the whole of the water; it was the same as that which now comes up from the gully-holes in the streets; the whole river was for the time a real sewer. (Faraday 1855; Also see Halliday 2013)

Clearly, passage of new health laws was not, in and of itself, going to be adequate to deal with London's waste disposal issues. (Only a few years later, Paris experienced a very similar event, later to become known as the Great Stink of Paris [Barnes 2006].)

British efforts to improve waste management facilities were reflected in similar legislation in the United States. In 1849, the Massachusetts state legislature created a three-man commission to conduct a "Sanitary Survey of the State." The chair of the commission was Boston historian and bookseller Lemuel Shattuck. A year later, the commission presented its "Report of the Sanitary Commission of Massachusetts." The report contained many of the same observations and criticisms of waste management systems in the United States as Chadwick had reported a decade earlier. The report consisted of four parts, the first two of which provided a review of the "sanitary movement" in Great Britain, other European countries, and the United States. The third section offered fifty recommendations for improving the city of Boston's sanitary system, while the final section provided supporting arguments for the recommendations and offered a model public health law for the country ("The Shattuck Report" 1850).

A critical feature of the new age of sanitation was improved toilet systems. By the middle of the 19th century, the population of London, Manchester, and other large cities had begun to explode. The Industrial Revolution brought large numbers of men and women from the countryside into urban areas, placing stresses on the infrastructure in those regions. According to some estimates, as many as a hundred individuals might have been forced into using a single toilet in some cities, and urban sewage systems were simply overwhelmed by burgeoning population numbers ("A Brief History of the Flush Toilet" 2019).

This increase presented a new challenge for those concerned with better sanitation in the city. Instead of allowing individuals simply to use their backyards or empty lots nearby to relieve themselves, indoor plumbing became a necessity, not a luxury. It was under this pressure that the development of the modern flush toilet first began with an invention by Scottish watchmaker and inventor Alexander Cumming in 1775. Cumming invented the characteristic S-shaped pipe under a toilet seat that is an essential feature of the modern-day toilet's functioning. Further development in toilet technology continued over the next century, reaching a peak in 1861, when inventor Thomas Crapper designed a revised version of Cumming's model for use in several royal palaces throughout Great Britain. A turning point in this story occurred in 1848 with the adoption of the Metropolitan Sewers Act, which forbade the construction of new houses lacking indoor toilets (water closets) and connections to municipal sewer systems (Akala 2018, 233; Allen 2002; Eveleigh 2008).

At about the same time in Great Britain and the United States, inventors were beginning to think about alternate methods for dealing with the nations' waste disposal problems. That work came to fruition in Great Britain in 1874 with the construction of the world's first waste incinerator, called the *Destructor*. The device consisted of a large storage bin into which municipal wastes were deposited. They were then moved from the bin to a large furnace, where they were burned at high temperature. The heat thus produced was either vented to the atmosphere or used to generate electricity and power pumps that ran the area's sewer pumps. The devices became very popular, and more than 200 were built over the next forty years. Similar devices, called *cremators*, were introduced into the United States at about the same time and became equally popular. More than 700 of the devices were built in this country prior to World War I ("1874—Furnace Incinerator for Refuse at Nottingham, England" 2016; Herbert 2007, 16–18; Melosi 2005, 39–40).

By the end of the 19th century, interest in the use of recycling to deal with waste disposal problems began to become more popular. As noted above, some forms of recycling had always been available throughout human history. For example, in 1690, the first paper mill to be built in the United States, the Rittenhouse Mill in Philadelphia, relied exclusively on recycled materials, particularly rags and used paper, as raw materials in its operation (Whitten 2009). A much later development came in 1897 when New York City instituted a program of "picking yards" in which all manner of waste products, from paper, metals, and fibers to rubber and horse hair, were deposited and then separated by category for future reuse in new products ("Then and Now: The Evolution of Recycling" 2019; Zarin 1987). A similar movement was underway in Great Britain where, for example, the British Paper Company was established in 1890 to use recycled rags and other materials to make paper.

The raw materials were collected by various organizations, such as the Salvation Army and individual "rag and bone" collectors (Aulston 2010, Appendix A).

As with composting and recycling, landfills continued to be a common form of waste management at the end of the 19th century. By the 1920s, however, a new approach to the technology had been adopted: sanitary landfills. Well, not new, since the technology had been known at least since its use in Knossos in the third millennium BCE, but new in the sense in that it superseded a far more common form of burying wastes, the simple landfill. The sanitary landfill differed from that technology by having alternate layers of refuse and clean dirt. This improvement reduced the problem of odor, rodents, leaching of materials, and other drawbacks of the simply dump-like landfill.

Authorities differ considerably as to the first appearance of the sanitary landfill, with estimates ranging from 1912 to the early 1920s. The United Nations Terminology Database, UNTERM, for example, makes the common claim that "[t]he method was introduced in England in 1912 (where it is called controlled tipping)" ("Sanitary Landfill" 2019). Other sources say that the technology was first installed in the city of Bradford, England, in 1915, 1916, or 1920 (Aston 1999, 110; Zimring and Rathje 2012, 471) or at an unspecified location "in the 1920s" (Melosi 2002). What is not at dispute is the rapid rise in interest and use of the sanitary landfill, first in Great Britain, and not much later, in the United States. Controlled tipping soon became the most popular form of waste management in Great Britain, with 834 out of 1,226 local authorities in England and Wales depending on this technology by 1968. A similar growth was observed in the United States, after the nation's first sanitary landfill opened in Fresno, California, in 1937. By 1965, the nation's total landfill capacity had grown to 85.2 million tons and, by 1990, to 145.3 million tons of refuse ("Municipal Solid Waste Generation, Recycling, and Disposal in the United States: Facts and Figures for 2012" 2012, table 3, p. 9).

Waste Management Legislation

The Public Health Acts of 1848 and 1875 in Great Britain might seem to have marked the beginning of a new era in solid waste legislation in that country and, perhaps, elsewhere. Such was by no means the case. For nearly a century after these bills were passed, few laws dealing with solid waste issues were adopted in either Great Britain or the United States. As one observer has noted,

> Politicians and the UK Government did not regard environmental protection as greatly important, except when forced by overwhelming evidence to act on the deaths caused in London, or other big cities, by air pollution or when particular chemicals were proven to be deadly even though there were equally effective alternate technologies available. Lawmakers had other, greater priorities for them, for the public, and for business. (Jones and Tansey 2015, xvi)

Such was also the case in the United States. The earliest environmental law adopted in the United States was the Rivers and Harbors Act, passed in 1882 and then reauthorized in its modern form in 1899. The law made it a felony to discharge refuse of any kind into any navigable body of water in the United States. Although since superseded by other legislation, the act is still used as a legal precedent for environmental actions in the country. The next piece of environmental legislation of any consequence was not passed for half a century, the Federal Water Pollution Control Act of 1948. That act also dealt exclusively and specifically with the dumping of wastes into the nation's waterways.

The first legislation dealing with solid wastes was the Solid Waste Disposal Act of 1965 (SWDA). That act provided federal funds to help states deal with issues of solid waste disposal, such as waste collection, transportation, recovery, and disposal. No state laws dealing with solid waste disposal existed at the time, and only ten full-time employees existed nationwide

(Hickman 2000; the text of the Solid Waste Disposal Act can be found at https://legcounsel.house.gov/Comps/Solid%20Waste%20Disposal%20Act.pdf, accessed on June 26, 2019).

The federal government found rather quickly that the SWDA was inadequate to deal with the growing problem of municipal and industrial solid wastes in the United States. Individuals and industries were generating wastes at a rate never seen before or imagined by experts in the field. The volume of solid wastes increased by 60 percent in the decade from 1950 to 1960, with an estimated 4 million chemicals included in those wastes. The *New York Times* opined in a 1969 editorial that "an avalanche of waste and waste disposal problems is building up around the nation's major cities in an impending emergency that may parallel the existing crises in air and water" ("25 Years of RCRA: Building on Our Past to Protect Our Future" 2002, 1). By 1960, the total mass of solid wastes generated in the nation had reached a total of more than 88 million tons, a figure that was to reach about 121 million tons a decade later ("Municipal Solid Waste Generation, Recycling, and Disposal in the United States: Facts and Figures for 2012" 2012, table 3, p. 9).

Congress' first response to this situation was to amend the SWDA to shift the emphasis from waste disposal to recycling and energy generation from solid wastes. That approach was expressed in the Resource Recovery Act of 1970 (RRA). One of the important sections of the law directed the Department of Health, Education, and Welfare to conduct a study of hazardous waste generation and disposal in the country. The result of that study was one of the nation's first warnings about the extent and threat posed by hazardous wastes to citizens. In its 1974 summary of research on hazardous wastes to the Congress, the Environmental Protection Agency (EPA) said,

> The management of the Nation's hazardous residues—toxic, chemical, biological, radioactive, flammable, and explosive wastes—is generally inadequate; numerous case studies demonstrate that public health and welfare are unnecessarily threatened by the uncontrolled discharge of

such waste materials into the environment. ("Report to Congress: Disposal of Hazardous Wastes" 1974, ix)

SWDA and RRA were adopted at a time when the United States was first beginning to recognize the environmental problems that had been developing—air, water, and solid waste pollution—since the end of World War II. That realization reached its peak in 1970 with the creation of the U.S. Environmental Protection Agency. The decade following that action was marked by the adoption of further, more extensive legislation on all aspects of the nation's pollution problems. In the area of solid wastes, the signature action was the passage in 1976 of another amendment to the SWDA, called the Resource Conservation and Recovery Act (RCRA). This act has been so important that it is now often thought of as providing the guiding principles for the nation's philosophy and practice on the treatment of solid wastes. It had four primary goals:

- Protecting human health and the environment from the potential hazards of waste disposal.
- Conserving energy and natural resources.
- Reducing the amount of waste generated.
- Ensuring that wastes are managed in an environmentally-sound manner. ("EPA History: Resource Conservation and Recovery Act" 2018; Also see "25 Years of RCRA: Building on Our Past to Protect Our Future" 2002, 2)

RCRA identifies three distinct programs in the area of solid waste management: nonhazardous wastes (Subtitle D), hazardous wastes (Subtitle C), and underground storage (Subchapter 1). These topics are currently covered under nine discrete programs: Hazardous Waste, Used Oil, Universal Wastes, Mixed Wastes, Land Disposal, Hazardous Waste Injection, Hazardous Wastes Imports/Exports, Permitting Program, and Underground Storage Tanks ("Resource Conservation and Recovery Act (RCRA) Compliance Monitoring" 2019; a guide to sections of the law dealing with specific solid waste issues is available at https://www.epa.gov/rcra/resource-conservation-and-recovery-act-rcra-regulations#nonhaz, accessed on June 26, 2019).

A second piece of legislation with maximum consequences for solid waste issues was the Comprehensive Environmental Response, Compensation, and Liability Act (CERCLA) of 1980, commonly known as Superfund. The act was passed in response to accumulating evidence of the industry's tendency to carelessly dispose of waste products without having or adhering to any specific plan for protecting the environment and human health. The program consists of two primary sections, the first of which is locating and identifying hazardous waste sites and the parties responsible for their creation and operation. The second phase involves programs for ending waste disposal at these sites, removing waste products located at the sites, and adopting programs of remediation to restore those sites to a safe condition.

A key element in the Superfund program is the National Priorities List (NPL), a list of all hazardous waste sites in the United States and its territories. As of late 2019, there were 1,187 sites on the NPL, forty-five that had been proposed for listing and 396 that had been deleted from the list (usually because of a successful cleanup operation). In addition to this status list, the EPA maintains a "milestone" list that reports the number of sites that are moving toward completion. In late 2019, fifty-two sites qualified for "partial deletion" status and 1,127 for "construction completion" status (meaning that all necessary physical work had been completed). These lists are constantly changing over time, trends for which can be found in table 5.6 (for an overview of the Superfund program, see https://www.epa.gov/superfund).

The four decades since the adoption of RCRA and CERCLA have seen a plethora of legislation on solid waste management, far more than can be adequately covered in this book. Some of the most important of these actions, however, are the following:

- The Used Oil Recycling Act of 1980 focused on the labeling and control of recycled oil products ("H.R.7833—Used Oil Recycling Act of 1980" 1980).

- The Hazardous and Solid Waste Amendments of 1984 expanded the scope of the RCRA by strengthening the regulation of entities generating small quantities of hazardous wastes and creating new requirements for other hazardous waste generators (Rosbe and Gulley 1984).
- The Ocean Dumping Ban Act of 1988 was passed because of a series of beach closings around the country due to solid waste floating ashore after being dumped in the open seas. The act banned the release of sewage sludge and industrial wastes in the oceans except for a small number of limited exceptions (Bortman 2016).
- The Medical Waste Tracking Act of 1988 was passed because of a spate of incidents in which medical wastes were being washed up on beaches on the nation's east coast. The act defined medical wastes and prescribed a *cradle-to-grave* program for their disposal and control (*Cradle-to-grave* is a term used in waste management that describes everything that happens to a product from the time it is created [the cradle] until it is finally disposed of [the *grave*].) ("Medical Waste Tracking Act of 1988" 2016).
- The Pollution Prevention Act of 1990 (P2) strengthened the emphasis on the reduction of waste production, use of recycling wherever possible, and turning to disposal of waste only as a matter of "last resort" (Pollution Prevention 2020).
- The Indian Lands Open Dump Cleanup Act of 1994 was instituted when researchers found that only two of more than 600 dumps in Indian territories met existing EPA waste management standards. The act provided funds for inventorying and closing noncompliant facilities (Gee 2007).
- The Mercury-Containing and Rechargeable Battery Management Act of 1996 aimed to phase out all use of mercury batteries in the country and to require labeling of other types of batteries that would encourage their safe use, recycling, and disposal ("Mercury-Containing and Rechargeable Battery Management Act" 1996).

- The Hazardous Waste Electronic Manifest Establishment Act of 2012 addressed the growing problem of electronic wastes (so-called e-wastes), which often contain a number and variety of hazardous materials. The bill established a cradle-to-grave system for tracking products that fall into this category of wastes.

(A very useful summary of all solid waste laws can be found at Bearden et al. 2013.)

Current Status of Solid Wastes in the United States

> Improper MSW disposal and management causes all types of pollution: air, soil, and water. Indiscriminate dumping of wastes contaminates surface and ground water supplies. In urban areas, MSW clogs drains, creating stagnant water for insect breeding and floods during rainy seasons. Uncontrolled burning of MSW and improper incineration contributes significantly to urban air pollution. (Alam and Ahmade 2013)

Statements like these are common in the literature about solid waste management. One legitimate question might then be this: Exactly *how serious* and *how dangerous* a threat do solid wastes pose today in the United States and in other parts of the world? One look at current data and data trends can help answer that question.

A total of 262.43 million tons of solid wastes were produced in the United States in 2015 (the most recent year for which data are available). That number represents an increase of about 200 percent over the total in 1960. Solid waste generation has increased every year from 1960 to 2015. (The EPA first began collecting data on solid waste generation and its ultimate fate in 1959. During 1960–2015, the U.S. population grew from 180 million to 320.9 million.) Although a substantial increase, another piece of data provides a more optimistic outlook: the amount of solid waste produced annually per person in the country. In 1960, that number amounted to 2.68 pounds per person per day, and in 2015, 4.48 pounds per person per day—an increase of only about 67 percent. (Unless otherwise noted, all data presented here come from "Advancing Sustainable Materials Management: 2015 Fact Sheet" [2018] or "National Overview: Facts and Figures on Materials, Wastes and Recycling" [2018]. For more detail on some of these topics, see tables 5.1–5.5.)

The United States currently ranks first among all the nations in the world in terms of the amount of MSWs produced. According to a 2012 report by the Urban Development and Local Government Unit of the World Bank, the United States was producing 624,700 metric tons of wastes per day (2.58 kilograms per person per day), followed by China (520,548 metric tons per day, 1.02 kilograms per person per day), Brazil (149,096 metric tons per day, 1.03 kilograms per person per day), and Japan (144,466 metric tons per day, 1.71 kilograms per person per day). Countries with the largest per-day disposal rate of solid wastes were the relatively small nations of Kuwait (5.72 kilograms per person per day), Antigua and Barbuda (5.50 kilograms per person per day), and Saint Kitts and Nevis (5.45 kilograms per person per day). The World Bank report predicted some important changes in these patterns over the next decade, with China becoming the largest producer in the world of MSWs (1,397,755 metric tons per day), followed by the United States (701,709 metric tons per day), India (376,639 metric tons per day), and Brazil (330,960 metric tons per day) (Hoornweg and Bhada-Tata 2012, Annex J, p. 80).

The largest component of solid wastes generated in 2015 was paper and paperboard, which accounted for 25.9 percent of all wastes produced. The next most important components of wastes were food wastes (15.1 percent), yard trimmings (13.2 percent), and plastics (13.1 percent). Other materials accounting for less than 10 percent of all solid wastes were metals (9.1 percent), wood (6.2 percent), textiles (6.1 percent), glass (4.4 percent), rubber and leather (3.2 percent), miscellaneous inorganic products (1.5 percent), and all other materials (2.0 percent).

Solid wastes in the United States are disposed of in one of four primary methods: landfills, recycling, combustion for energy, and composting. The most common of these methods in 2015 was landfills, which accounted for 52.5 percent of wastes generated. Another 25.8 percent was recycled, 12.8 percent combusted for energy, and 8.9 percent composted. As table 1.1 shows, these numbers changed significantly in the period between 1960 and 2015. The popularity of landfills increased to a maximum in 1980–1990, when disposal rates with this resource reached 3.2 pounds per person per day. Since that time, they have fallen off to 2.3 pounds per person per day, the lowest rate since the EPA started keeping these records. By contrast, the three other disposal methods increased in popularity. The rates for recycling and combustion for energy have both increased by about 600 percent (0.2–1.2 pounds per person per day and less than 0.1 to 0.6 pounds per person per day,

Table 1.1 Waste Disposal Patterns, 1960–2015

Millions of tons annually									
Activity	**1960**	**1970**	**1980**	**1990**	**2000**	**2005**	**2010**	**2014**	**2015**
Landfill	82.5	112.6	134.3	145.3	140.3	142.2	136.3	136.2	137.7
Recycling	5.6	8.0	14.5	29.0	53.0	59.2	65.3	66.6	67.8
Combustion	0.0	0.5	2.8	29.8	33.7	31.7	29.3	33.2	33.5
Composting	neg.	neg.	neg.	4.2	16.5	20.6	20.2	23.0	23.4
Pounds per Person per Day									
Activity	**1960**	**1970**	**1980**	**1990**	**2000**	**2005**	**2010**	**2014**	**2015**
Landfill	2.5	3.1	3.2	3.2	2.7	2.6	2.4	2.3	2.3
Recycling	0.2	0.2	0.4	0.6	1.0	1.1	1.1	1.1	1.2
Combustion	0.0	neg.	0.1	0.7	0.7	0.6	0.5	0.6	0.6
Composting	neg.	neg.	neg.	0.1	0.3	0.4	0.4	0.4	0.4
U.S. Population*	180.0	204.0	227.3	249.9	281.4	296.4	309.1	318.9	320.9

neg. = negligible.

*in millions.

Source: "Advancing Sustainable Materials Management: 2015 Fact Sheet." 2018. Tables 2 and 3, pp. 5–6. Environmental Protection Agency. https://www.epa.gov/sites/production/files/2018-07/documents/2015_smm_msw_factsheet_07242018_fnl_508_002.pdf. Accessed on June 28, 2019.

respectively), while composting has increased in popularity by about 400 percent.

The composition of solid wastes has also undergo a change in the period from 1960 to 2015. By far, the largest material found in solid wastes in 1960 was paper and paperboard, making up about 34 percent of all solid wastes in the country. The second most common component of solid wastes was metals of all kinds (12.3 percent) followed by glass (7.6 percent). By 2015, the numbers for all three materials had decreased to 25.9 percent, 9.1 percent, and 4.4 percent, respectively. The most remarkable change that occurred during the half century was the presence of plastics in solid wastes. In 1960, those products made up only 0.4 percent of all wastes, a number that jumped to 13.1 percent in 2015, becoming the second-largest component of solid wastes in the country. (See table 5.2 for more details on these data.)

One of the most dramatic changes in solid waste disposal patterns in the United States between 1960 and 2015 is the use of recycling as a method of disposal. As shown in table 1.2, recycling rates for every major type of solid waste has increased dramatically over that period of time. Perhaps the most

Table 1.2 Recycling Rates for Various Materials, 1970–2015

Material	1970	1980	1990	2000	2014	2015
Paper and paperboard	15%	21%	28%	43%	65%	67%
Glass	1%	5%	20%	23%	26%	26%
Metals	4%	8%	24%	35%	35%	34%
Plastics	neg.	<1%	2%	6%	10%	9%
Yard trimmings	neg.	neg.	12%	52%	61%	61%
Selected consumer electronics				10%	42%	40%
Lead-acid batteries	76%	70%	97%	93%	99%	99%

neg. = negligible.

Source: "National Overview: Facts and Figures on Materials, Wastes and Recycling." 2018. Environmental Protection Agency. https://www.epa.gov/facts-and-figures-about-materials-waste-and-recycling/national-overview-facts-and-figures-materials#Generation. Accessed on June 28, 2019.

dramatic single pattern has been that for lead storage batteries, nearly 100 percent of which are now routinely recycled. The one disappointing feature of the chart, however, is the low rate of plastic recycling. That rate has increased from essentially zero in 1970 to about 10 percent today, still a remarkably low rate for such a serious pollutant.

A form of solid waste that is not included in discussions of MSWs is the so-called *construction and demolition debris* (C&D). These are waste materials produced during the construction of buildings, roads, bridges, and other structures. They include Portland cement concrete (67 percent of all C&D wastes for 2013), asphalt concrete (18 percent), wood products (8 percent), asphalt shingles, drywall and plasters, (2 percent each), and steel (1 percent). The vast majority of this debris is produced during the demolition processes (505.9 million tons in 2013) compared to that produced during the construction processes themselves (24.4 million tons). (See "Advancing Sustainable Materials Management: 2015 Fact Sheet" 2018, 17–19.)

Many researchers have attempted to determine the precise effects of solid wastes and management methods on the environment and human health. Chapter 2 includes a discussion of some findings from this research. One topic to which considerable attention has been paid recently is the effect of waste management technologies, recycling in particular, on climate change trends. Table 1.3 summarizes one line of research designed to estimate the number of cars that could be removed from U.S. highways annually because of solid waste recovery activities. The table shows, for example, that the recovery of 2.47 million tons of wood wastes in 2013 produced a savings of 3.3 million tons of greenhouse gases (GHG), which, in turn, was equivalent to removing 798,000 cars from American roads and highways in that year.

Integrated Solid Waste Management

For most of human history, solid waste management was a relatively unscientific, somewhat disorganized, and often inefficient

Table 1.3 Greenhouse Gas and Number of Cars Savings Associated with Recovery of Certain Types of Wastes in 2013

Material	Weight Recovered*	GHG Equivalent (CO_2)*	Number of Cars Saved
Paper and paperboard	43	149	31 million
Glass	3.2	1	210,000
Metals			
Steel	5.8	9.5	2 million
Aluminum	0.7	6.4	1.3 million
Other nonferrous metals	1.37	5.9	1.2 million
Total metals	7.87	21.8	4.5 million
Plastics	3	3.6	760,000
Rubber and leather	1.24	0.6	127,000
Textiles	2.3	5.8	1.2 million
Wood	2.47	3.8	798,000
Other wastes			
Food and other organics	1.84	1.7	308,000
Yard trimmings	20.6	1.04	220,000

*in millions of tons.

Source: "Advancing Sustainable Materials Management: 2015 Fact Sheet." 2018. Table 5, p. 15. Environmental Protection Agency. https://www.epa.gov/sites/production/files/2015 09/documents/2013_advncng_smm_fs.pdf. Accessed on June 29, 2019.

method for dealing with a community's solid waste problems. Some agency might construct a landfill in which wastes were deposited. The landfill was often no more than an open pit into which wastes were placed and then covered with "clean" soil. Over the years, the landfill would no longer be able to accept wastes, and the agency would move on to beginning a new landfill. Meanwhile, the old landfill just sat there, with at least some of its wastes escaping into the atmosphere, the surrounding land, or the water table.

In this cycle of events, noxious products were relatively easily spread to surrounding areas, producing unsightly conditions, unpleasant odors, and threats to human and other forms of animal health. Some of the health problems commonly associated with poorly constructed and managed landfills are skin and blood infections; eye and respiratory diseases; and a host of chronic disorders, such as hepatotoxicity (liver damage), nephrotoxicity (kidney damage), pulmonary toxicity (lung disease), neurotoxicity (nerve damage), and immunotoxicity (damage to the immune system) (Sridevi et al. 2012). Landfills are also a breeding ground for many varieties of disease-carrying animals, such as flies, cockroaches, mice, rats, that transmit a variety of diseases including malaria, cholera, and dysentery.

The long-used practice of feeding wastes to pigs also has had its serious downside. Pathogens present in the wastes fed to swine may cause a variety of diseases, such as foot-and-mouth disease, brucellosis, tuberculosis, and gastroenteritis (Hickman and Eldredge 2016; "Risks Associated with Feeding Raw or Improperly Cooked Food Waste to Swine" 2018).

Incineration also had its drawbacks. Among the most obvious is the air pollution that reduces visibility, increases the spread of odor, and spreads its own variety of diseases. Studies have shown that the effluent from garbage incineration tends to include pollutants such as heavy metals like lead and mercury, respiratory irritants like sulfur dioxide and nitrogen oxides, and toxic chemicals like dioxins and furans. These irritants cause their own suite of health disorders, primarily a variety of lung disorders (*Waste Incineration and Public Health* 2000, chapter 5, https://www.ncbi.nlm.nih.gov/books/NBK233619/, accessed on July 2, 2019). The ash that remains after waste incineration also has its own set of health problems. This ash occurs in a variety of forms—namely, fly ash, bottom ash (slag), heat recovery ash, and air pollution residues. These products contain a host of harmful chemicals, the most dangerous of which are a group of products known as persistent organic products (POPs). The group consists of carcinogenic, teratogenic, and toxic substances such as polychlorinated biphenyls (PCBs), hexachlorobenzene (HCB), polychlorinated dibenzo-p-dioxins (PCDDs), and dibenzofurans (PCDFs). The last two groups are also known simply as *dioxins*. These substances have been found to produce measurable and harmful effects on individuals living even at a relatively distant location from waste incinerators ("After Incineration: The Toxic Ash Problem" 2005).

An Overview

The fundamental issue associated with all forms of waste disposal has been the tendency to think of them in isolation from other environmental and health consequences. For example, agencies that decided to use incineration as a primary waste disposal method usually gave little or no thought to short- and long-term changes that technology might have on the environment or on human health, as noted above. By the late 1940s, then, a handful of scholars began to think about waste management issues in a unitary way—that is, as a complete system in which all steps in the waste generation and disposal process were connected with each other. That line of thought has led to the rise of a more sophisticated form of solid waste management known as integrated solid waste management (ISWM). One definition of ISWM is "a complete waste reduction, collection, composting, recycling, and disposal system." "An efficient ISWM system," that definition continues, "considers how to reduce, reuse, recycle, and manage waste to protect human health and the natural environment. It involves evaluating local conditions and needs. Then choosing, mixing and applying the most suitable solid waste management activities according to the condition" (LeBlanc 2019).

The concept of ISWM includes three major concepts. The first is a melding of the basic elements of any solid waste management program: waste generation, disposal (landfills and incineration), and recycling and composting. That is, ISWM is based on waste management technologies that have been around for centuries, if not millennia, except that they are considered as parts of a grander scheme of waste management.

The second major concept is the consideration given to local conditions in the design of any solid waste management program. That is, no one "cookie-cutter" approach to solid waste management is appropriate for all situations. Instead, designers of such a program must take into consideration the specific geography, topography, demographics, economics, environmental features, racial and ethnic makeup, history, and other factors of a specific community in deciding how to develop the features of an ISWM for that community.

A third important concept is sustainability. A traditional solid waste management program is usually not sustainable. That is, its operation cannot continue in its basic form forever or, in many cases, even for a few decades. The philosophy of ISWM is to find adaptations that can be made in traditional methods at every stage of the solid waste management process to ensure that the system does not overwhelm the planet's ability to function efficiently ("Sustainable Materials Management: The Road Ahead" 2009).

The Details

Once some guiding principles for ISWM have been developed, questions remain as to how those principles can actually be put into practice to solve solid waste management problems. Specific procedures have been suggested for each stage of the solid waste management process. One of the best available plans for implementing ISWM principles in a practical way on a large scale is a program developed by the U.S. Army Public Health Command. The plan describes in detail the ways in which army installations can implement each of the four major steps in ISWM at their facilities. (See "Guide for Developing Integrated Solid Waste Management Plans at Army Installations" 2013.)

Generation

In some respects, the most important changes that ISWM programs need to make occur at the generation state: the United States and other nations are simply producing too much "stuff." Currently available waste disposal and management systems cannot keep up with the volume of paper, wood, metals, plastics, and other materials that are being generated every year. In the United

States, the amount of MSWs has increased from 88.120 million tons in 1960 to 262.430 million tons in 2015 (the last year for which data are available; "Materials Generated in the Municipal Waste Stream, 1960 to 2015" 2018, table 35, p. 46).

Imagine that you are the owner of the Super Plastico company. Your product is very popular worldwide, at least partly because it is so tough and long lasting. Customers find that your products last many years, and even many decades. They just never seem to wear out.

But a new problem has arisen for the company. Your product has been showing up in some unexpected and undesirable places, such as popular swimming beaches, in the middle of Lake Michigan, and in the stomach of farm animals. And they have been, and probably will continue to be in these places, for years, decades, or even centuries to come. As a responsible person concerned about the environment, none of this good news. What can Super Plastico do to deal with this problem?

Integrated social waste management to the rescue! The first and most important message you hear is that you can no longer stop thinking about your product when it comes off the production line and goes out the door to Wichita or Lima. You have to begin to think about how and where it might escape into the environment and what you can do to prevent those unpleasant events from happening. This process of monitoring the production of items is often called *source reduction*. Some suggestions the experts have given you are:

1. Reconsider the materials from which your product is made. Is there a plastic that breaks down more easily? Can you substitute a different material entirely, such as a light metal or carbon fiber product, which will decompose more easily?
2. Are there ways you can modify the system by which the product is made? Every manufacturing process results in some waste of materials, but some are more efficient than others. Is the one you are using the best one for reducing factory wastes?

3. Can you redesign the product to reduce the amount of material used in its production? Sometimes an extra piece can be taken out without affecting the efficiency of your product, and in doing so, you might save both money and future wastes.
4. Can you improve your packaging of the final product? Sometimes it's tempting to use four or five layers of packing material to prevent damage during transit. But maybe you can get along with two or three layers instead, resulting in a substantial savings on wastes.
5. Maybe you can find a way to reuse production materials in your plant. Sometimes it's easy just to throw away inexpensive materials and devices used in manufacturing a product. But you may be able to save yourself money and reduce waste by making those items last longer in the production process.
6. Consider focusing on waste production in your plant's operation, perhaps by offering educational programs for your employees. Help them understand what you already know: that one of the most important keys to reducing solid wastes is at the source from which products come.
7. Develop an auditing system for materials used in your manufacturing process. It can be helpful to have hard figures to look at to see how and where savings can be made that will reduce waste. (Many good resources on waste generation management are available on the internet. See, for example, Brucker 2017; Hupp 2007; Lawson 2018.)

Recycling and Composting

As part of an ISWM system, recycling and composting are the exact opposite of a dominant theme in many developed countries today. That theme is based on the concept of "use it up" and "throw it away." Many products are actually designed today with that concept in mind: buy a product, use it until it breaks down or wears out, and then just throw it away and buy the same or a similar product to replace it. Recycling involves the collection and separation of glass, plastic, paper, metal, and other types of wastes before reusing them in the production of new items. The process of recycling is often subdivided into three types of actions: upcycling, recycling, and downcycling. The three processes involve the conversion of a waste material into an object that has greater value, equal value, or less value than the original product being recycled.

Composting is a waste disposal procedure similar to recycling except that it involves the processing of organic materials, such as food wastes, into a form in which they can be used in the environment, such as food for farm animals or fertilizers. The most common materials used in compositing are yard trimmings, food scraps, and biosolids (organic matter obtained from sewage). A variety of mechanisms is now available for the conversion of these materials into products useful in agriculture and the dairying industry. (For an excellent overview of the status of composting in the United States today, see Platt, Goldstein, and Coker 2014. Also see the essay on composting by Lisa Perschke in chapter 3.)

For many experts, recycling has long been the best possible technology for dealing with wastes (compared to landfills and incineration). It conserves wastes for reuse rather than burying them in the ground or burning them for energy generation (or for no practical purpose at all). Since the early 2000s, a competing philosophy of waste management has developed: zero waste. The term *zero waste* refers to systems in which no wastes are produced at all. A common definition for the process is one that aims for "[t]he conservation of all resources by means of responsible production, consumption, reuse, and recovery of products, packaging, and materials without burning and with no discharges to land, water, or air that threaten the environment or human health" ("Zero Waste Definition" 2018). The zero waste philosophy rejects recycling as an essential part of the waste management process, one that is to be used only as a last resort, a way of avoiding the dumping of waste materials into some part of the natural environment. As one website explains, the goal of zero waste is "to completely eliminate waste, not manage it." ("Zero Waste Versus Recycling: What's the Difference?" n.d.)

Until zero waste becomes a practical reality, recycling and composting are likely to remain popular methods for the recovery of materials and objects that would otherwise find their way into an incinerator or landfill. According to the most recent data available, 262.4 million tons of MSWs were recycled in 2015. That is the largest number since the Environmental Protection Agency began keeping records in 1960. The largest fraction of recycled material was paper and paper products, which made up 25.9 percent of all recycled wastes, followed by plastics (13.1 percent), metals (9.1 percent), wood (6.2 percent), and textiles (6.1 percent). Meanwhile, 23.4 million tons of wastes were composted, 61.3 percent of which were yard trimmings, 29.8 percent were biosolids, and 5.3 percent were food scraps. This total was 450 percent greater than the total in 1990 ("Materials Recycled and Composted in the Municipal Waste Stream, 1960 to 2015" 2018, table 2).

One of the most common forms of recycling is container deposit laws, otherwise known as *bottle bills*. A container deposit law provides for a customer's paying a small down payment (often five or ten cents) on a bottle, can, jug, or other type of container at the time of purchasing a product. Upon returning that container, the customer then gets her or his deposit back. The first such bill was passed in the state of Oregon in 1971. The original deposit was five cents, later increased to its current ten cents. Today, ten states have such legislation: California, Connecticut, Hawaii, Iowa, Maine, Massachusetts, Michigan, New York, Oregon, and Vermont. Bottle bills tend to be quite successful. In Michigan, for example, between $294.4 million (1990) and $457.8 million (1998) has been collected in deposits every year since the adoption of the law in 1990.

The return rate for these deposits has been as high as 100.4 percent (1992) and as low as 91.2 percent (2017), but averaging about 97 percent during this seventeen-year period ("Michigan Bottle Deposit Law. Frequently Asked Questions" 2019). The success of bottle bills is also reflected in comparisons of recycling rates between states that do have bottle bills and those that do not. In its 2013 report on container recycling, the Container Recycling Institute noted that states with bottle bills had an 84 percent recycling rate for aluminum containers, while states without such laws had a recycling rate of 39 percent. Comparable rates for PET plastics and glass were 48 percent versus 20 percent and 65 percent versus 25 percent, respectively ("Bottled Up (2000–2010)—Beverage Container Recycling Stagnates Download" 2013). As might be expected, bottle bills come with their own set of disadvantages that make them difficult to pass and execute. For example, the recycling rate for containers in Michigan decreased from 96.1 percent in 2010 to 91.2 percent ("Michigan Bottle Deposit Law. Frequently Asked Questions" 2019). The reasons for a change such as this one and a discussion of problems accompanying bottle bills will be discussed in chapter 2.

Incineration

Probably the most certain method of waste *disposal* is incineration. In its simplest possible terms, *waste incineration* simply means burning up solid wastes. During the incineration process, those wastes are converted to gases, which escape into the atmosphere or are captured for use in other processes, and ash and solid products that remain after incineration is complete. At one time, and even today in some locations, incineration involved nothing more than setting fire to waste and letting them burn themselves out. Much more common now, however, is the capture of the heat energy produced during incineration. That energy can then be used directly, for the heating of large buildings, for example, or the generation of electricity.

Waste-to-energy plants are not equally popular in all parts of the world. In 2015, Japan burned 68 percent of all its MSWs, with recovery of the energy produced. It was followed by

Norway, with 54 percent of its solid wastes incinerated with energy recovery; Switzerland with 48 percent; and France with 35 percent. At the same time, 13 percent of all MSW in the United States was being treated by incineration with energy recovery ("How Waste-to-Energy Plants Work" 2019).

In the basic model of a waste-to-energy plant, trash is dumped into a collecting bin that connects with a combustion chamber. (See the diagram at "How Waste-to-Energy Plants Work" [2019] for a good illustration of this process.) The heat produced by combustion of the trash is then used to heat water in a boiler connected to the combustion chamber. The boiler is connected, in turn to a steam drum, where water is changed into steam. The steam is then passed to a steam turbine, where it is used to run a generator that produces electricity. In some cases, the heat that comes from the combustion chamber can be diverted directly into home, commercial, and industrial heating systems.

Some portion of the original trash delivery is not combustible. It is converted in the combustion chamber to ash, which is drawn off at the bottom of the chamber. The ash can then be separated into useful products, such as metals that can be recycled and waste products that are deposited into landfills.

Other types of waste-to-energy plants also exist. For example, pyrolysis plants use very hot combustion chambers (more than 400°C) where combustion occurs much more quickly. Gasifiers are plants that operate at an even higher temperature (550°C–1,500°C) in the absence of oxygen. Under these conditions, wastes are converting directly to combustible gases that can then be used for industrial operations. Generally speaking, these variations on waste-to-energy processes are not economically viable for very large operations. The basic model described above, then, is more commonly used for such systems (Stringfellow 2014).

Specialized incineration plants are also available for the disposal of hazardous wastes. The most common approach in such plants is the use of very high temperatures, usually in excess of

1,800°C. At these temperatures, most wastes are converted to harmless products such as water, carbon dioxide, nitrogen, and oxygen. Harmful products, such as sulfur dioxide and nitrogen oxides can be captured and converted into harmless products, such as sulfates and nitrates. Incineration reduces the volume of hazardous materials that must be disposed of. Those not lost as gaseous products end up as ash, which can be more inexpensively disposed of in landfills than the original materials themselves (Karstensen 2019).

Landfills

Open dumps and landfills are, of course, among the oldest waste disposal mechanisms known to humans. They are also among the least effective waste disposal systems for protecting the environment and human health. As noted above, they are generally very inefficient for preventing all types of solid wastes and their products from escaping into the surrounding ground, groundwater, and air. For this reason, a considerable amount of research has gone into the development of more efficient landfills. These landfills are known as *sanitary* or *secure landfills*.

One essential component of a sanitary landfill is a lining that prevents waste materials from escaping into the surrounding ground. The lining generally consists of a thick plastic sheet covered with a clay lining. The landfill also contains a system for collecting leachates, liquids released from the solid wastes stored in the landfill. A system of pipes sunk into the wastes permits the collection of noxious gases released from the wastes. The most common of these waste gases, methane, may be burned to produce electrical energy or heat.

Another type of sanitary landfill is called a secure landfill. This system is designed for storage of toxic or hazardous wastes, such as those produced by various types of industrial operations. The basic elements of a secure landfill are the same as those for a sanitary landfill. But parts of the system are strengthened to ensure that waste products do not escape into the environment. For example, the thickness of the linings is increased, and a monitoring system may be included or made more sensitive to leaks. Industrial wastes are also stored in barrels or other containers to further reduce leakage into the environment ("Secure Landfill Disposal," n.d.; for a diagram of a secure landfill, see "Waste Disposal and Recycling. Reduce, Reuse, Recycle" 2019, slide #83).

Special Problems

The description of solid waste disposal provided here focuses especially on municipal wastes. Other types of wastes, such as nuclear, electronic, and universal wastes, generally require specialized methods of storage, transportation, and treatment. For example, nuclear wastes must usually be packed very carefully for transportation from one site to another and may even have to travel routes that avoid populated areas or busy highways. They generally have to be buried with recognition of the fact that they will remain harmful to humans for hundreds of year, or much longer. Electronic wastes pose special disposal problems also because the very small components of which they are made are often harmful to human health, as well as very costly and worthy of recovery. These systems and special problems are discussed in more detail in chapter 2.

References

"Advancing Sustainable Materials Management: 2015 Fact Sheet." 2018. Environmental Protection Agency. https://www.epa.gov/sites/production/files/2018-07/documents/2015_smm_msw_factsheet_07242018_fnl_508_002.pdf. Accessed on June 28, 2019.

"After Incineration: The Toxic Ash Problem." 2005. IPEN Dioxin, PCBs and Waste Working Group. https://ipen.org/sites/default/files/documents/After_incineration_the_toxic_ash_problem_2015.pdf. Accessed on July 2, 2019.

Akala, Jethron Ayumba. 2018. "In the Technological Footprints of Urbanity: A Socio-political History of Water and Sanitation in Nairobi, 1899–2015." http://tuprints.ulb.tu-darmstadt.de/8550/1/Jethron%20Ayumba%20Akala.pdf. Accessed on June 30, 2019.

Alam, Pervez, and Kafeel Ahmade. 2013. "Impact of Solid Waste on Health and the Environment." *International Journal of Sustainable Development and Green Economics*. 2: 165–168. https://pdfs.semanticscholar.org/2ae9/675a58adb025fb799703750cd477ca838bab.pdf. Accessed on June 27, 2019.

Allen, Michelle. 2002. "From Cesspool to Sewer: Sanitary Reform and the Rhetoric of Resistance, 1848–1880." *Victorian Literature and Culture*. 30(2): 383–402.

Amick, Daniel S. 2014. "Reflection on the Origins of Recycling: A Paleolithic Perspective." *Lithic Technology*. 39(1): 64–69. https://www.academia.edu/6076839/Reflections_on_the_Origin_of_Paleolithic_Recycling_A_Paleolithic_Perspective. Accessed on June 15, 2019.

Andersson, Magnus, et al. 2004. *Stone Age Scania: Significant Places Dug and Read by Contract Archaeology*. Lund, Sweden: National Heritage Board. Archaeological Excavations Department. https://www.academia.edu/29089411/Stone_Age_Scania._Significant_places_dug_and_read_by_contract_archaeology. Accessed on June 15, 2019.

Antoniou, Georgios P., et al. 2016. "Evolution of Toilets Worldwide through the Millennia." *Sustainability*. 8(8): 779. https://pdfs.semanticscholar.org/01b3/d3678a058f3f7e6daac692d090ce8a60cea8.pdf. Accessed on June 21, 2019.

Aston, Robert Lee. 1999. *The Legal, Engineering, Environmental and Social Perspectives of Surface Mining Law and Reclamation by Landfilling Getting Maximum Yield from Surface Mines*. London: Imperial College Press.

Aulston, Nichola Jay. 2010. "History and Current Status of Waste Management in the United States." https://www.academia.edu/387650/History_and_Current_State_of_Waste_Management_in_the_United_States. Accessed on June 25, 2019.

Barnes, David S. 2006. *The Great Stink of Paris and the Nineteenth-Century Struggle against Filth and Germs.* Baltimore: Johns Hopkins University Press.

Bearden, David M., et al. 2013. "Environmental Laws: Summaries of Major Statutes Administered by the Environmental Protection Agency." Congressional Research Service. https://fas.org/sgp/crs/misc/RL30798.pdf. Accessed on June 27, 2019.

Bortman, Marci L. 2016. "Ocean Dumping Ban Act of 1988." Encyclopedia.com. https://www.encyclopedia.com/environment/encyclopedias-almanacs-transcripts-and-maps/ocean-dumping-ban-act-1988. Accessed on June 27, 2019.

"Bottled Up (2000–2010)—Beverage Container Recycling Stagnates Download." 2013. Container Recycling Institute. http://www.container-recycling.org/index.php?option=com_content&view=article&id=394. Accessed on July 5, 2019.

Bradley, Mark, ed. 2012. *Rome, Pollution, and Propriety: Dirt, Disease, and Hygiene in the Eternal City from Antiquity to Modernity.* Cambridge, UK: Cambridge University Press.

"A Brief History of the Flush Toilet." 2019. The British Association of Urological Surgeons. https://www.baus.org.uk/museum/164/the_flush_toilet. Accessed on June 30, 2019.

Brucker, Drew. 2017. "18 Waste Reduction Tips for Small Businesses." Rubicon. https://www.rubiconglobal.com/blog-waste-tips-small-businesses/. Accessed on July 3, 2019.

"Campaign for Public Health (Summary)." 2019. Learning. Dreamers and Dissenters. British Library. https://www.bl.uk/learning/histcitizen/21cc/publichealth/summary/publichealthcampaign.html. Accessed on June 24, 2019.

"The 1848 Public Health Act." 2019. UK Parliament. https://www.parliament.uk/about/living-heritage/transformingsociety/towncountry/towns/tyne-and-wear-case-study/about-the-group/public-administration/the-1848-public-health-act/. Accessed on June 24, 2019.

"1874—Furnace Incinerator for Refuse at Nottingham, England." 2016. Historia Sanitaria. https://www.wiki.sanitarc.si/1874-fryer-builds-furnace-incinerator-refuse-nottingham-england/. Accessed on June 25, 2019.

"EPA History: Resource Conservation and Recovery Act." 2018. Environmental Protection Agency. https://www.epa.gov/history/epa-history-resource-conservation-and-recovery-act. Accessed on June 26, 2019.

Eveleigh, David J. 2008. *Privies and Water Closets*. Oxford, UK: Shire.

"Exhibition: Mount of Amphorae." 1997. Wayback Machine. https://web.archive.org/web/20050302200446/http://ceipac.gh.ub.es/MOSTRA/u_expo.htm. Accessed on June 20, 2019.

Faraday, Michael. 1855. "Observations on the Filth of the Thames, Contained in a Letter Addressed to the Editor of 'The Times' Newspaper, by Professor Faraday." Chem Team. https://www.chemteam.info/Chem-History/Faraday-Letter.html. Accessed on July 5, 2019.

"From Waste to Resource: An Abstract of '2006 World Waste Survey.'" 2016. Cyclope. Veolia Environmental Services. http://81.47.175.201/flagship/attachments/waste_resource.pdf. Accessed on June 15, 2019.

Gee, Randy. 2007. "Cherokee Nation/Inter-Tribal Environmental Council Open Dump Data Exchange." Cherokee Nation Environmental Programs. https://www.exchangenetwork.net/meetings-and-events/2007-en-national-meeting/. Accessed on June 27, 2019.

"Geoscience for America's Critical Needs." 2016. American Geosciences Institute. https://www.americangeosciences

.org/sites/default/files/AGI_GeoscienceForAmericasCritical Needs_102315_WebRes.pdf. Accessed on June 15, 2019.

Grattan, J. P., et al. 2016. "The First Polluted River? Repeated Copper Contamination of Fluvial Sediments Associated with Late Neolithic Human Activity in Southern Jordan." *Science of the Total Environment*. 573: 247–257.

"Guide for Developing Integrated Solid Waste Management Plans at Army Installations." 2013. U.S. Army Public Health Command. https://phc.amedd.army.mil/PHC%20 Resource%20Library/FINAL_TG_197_November_2013 .pdf. Accessed on July 3, 2019.

"H.R.7833—Used Oil Recycling Act of 1980." 1980. Congress.gov. https://www.congress.gov/bill/96th-congress/ house-bill/7833. Accessed on June 27, 2019.

Halliday, Stephen. 2013. *The Great Stink of London: Sir Joseph Bazalgette and the Cleansing of the Victorian Metropolis*. New York: History Press.

Havlíček, Filip. 2015. "Waste Management in Hunter-Gatherer Communities." *Journal of Landscape Ecology*. 8(2): 47–59. https://content.sciendo.com/view/journals/ jlecol/8/2/article-p47.xml. Accessed on June 15, 2019.

Havlíček, Filip, and Kuča Martin. 2017a. "Waste Management in Bronze Age Europe." *Journal of Landscape Ecology*. 10(1): 35–43. https://www.degruyter.com/downloadpdf/j/ jlecol.2017.10.issue-1/jlecol-2017-0008/jlecol-2017-0008 .pdf. Accessed on June 15, 2019.

Havlíček, Filip, and Kuča Martin. 2017b. "Waste Management at the End of the Stone Age." *Journal of Landscape Ecology*. 10(1): 44–57. https://content.sciendo.com/view/journals/ jlecol/10/1/article-p44.xml. Accessed on June 15, 2019.

Havlíček, Filip, and Miroslav Morcinek. 2016. "Waste and Pollution in the Ancient Roman Empire." *Journal of Landscape Ecology*. 9(3): 33–49. https://www.degruyter.com/ downloadpdf/j/jlecol.2016.9.issue-3/jlecol-2016-0013/ jlecol-2016-0013.pdf. Accessed on June 20, 2019.

Havlíček, Filip, Adéla Pokorná, and Jakub Zálešák. 2017. "Waste Management and Attitudes towards Cleanliness in Medieval Central Europe." *Journal of Landscape Ecology*. 10(3): 266–287. https://www.degruyter.com/downloadpdf/j/jlecol.2017.10.issue-3/jlecol-2017-0005/jlecol-2017-0005.pdf. Accessed June 21, 2019.

Herbert, Lewis. 2007. "Destructors—The Disposal Route of Choice 100 Years Ago." *Centenary History of Waste and Waste Managers in London and South East England*. https://www.ciwm.co.uk/Custom/BSIDocumentSelector/Pages/DocumentViewer.aspx?id=QoR7FzWBtitMKLGdXnS8mUgJfkM0vi6KMAYwUqgqau3ztZeoed%252bsdmKIqDzPOm8yAXgBZR%252fn1fYhL%252bTNdjUq9g2xwY63C2g8GcAQQyfpf3SImIrrED%252bTfsUM91bKsogr. Accessed on June 25, 2019.

Hickman, H. Lanier. 2000. "A Brief History of Solid Waste Management in the US, 1950 to 2000—Part 5a." Forester Media. https://www.foresternetwork.com/msw-management/waste/article/13000553/a-brief-history-of-solid-waste-management-in-the-us-1950-to-2000-part-5a. Accessed on June 26, 2019.

Hickman, H. Lanier, Jr., and Richard W. Eldredge. 2016. "A Brief History of Solid Waste Management in the US During the Last 50 Years—Part 1." Forester Media. https://www.foresternetwork.com/msw-management/article/13023338/a-brief-history-of-solid-waste-management-in-the-us-during-the-last-50-years-part-1. Accessed on July 2, 2019.

Hoornweg, Daniel, and Perinaz Bhada-Tata. 2012. "What a Waste. A Global Review of Solid Waste Management." Urban Development and Local Government Unit. World Bank. http://documents.worldbank.org/curated/en/302341468126264791/pdf/68135-REVISED-What-a-Waste-2012-Final-updated.pdf. Accessed on June 28, 2019.

"How Waste-to-Energy Plants Work." 2019. U.S. Energy Information Administration. https://www.eia.gov/energy explained/?page=biomass_waste_to_energy#tab2. Accessed on July 5, 2019.

Hupp, Jessica. 2007. "Working Green: 50 Tips to Reduce Your Office's Waste." Science Alert. https://energimeuniversity .org/working-green-50-tips-to-reduce-your-offices-waste/. Accessed on July 3, 2019.

Jones, E. M., and E. M. Tansey, eds. 2015. "The Development of Waste Management in the UK c.1960–c.2000." http:// www.histmodbiomed.org/sites/default/files/W56LoRes.pdf. Accessed on June 26, 2019.

Kahn, Lloyd, Blair Allen, and Julie Jones. 2007. *The Septic System Owner's Manual.* Bolinas, CA: Shelter Publications.

Karstensen, Kåre Helge. 2019. "Environmentally Sound Destruction of POPs." SlidePlayer. https://slideplayer.com/ slide/5101819/. Accessed on July 6, 2019.

Klick, Jacques. n.d. "The Effects of the Black Death on the Life, Culture and Mentality of Medieval People." Academia. https://www.academia.edu/7384340/The_Effects_of_the_ Black_Death_on_the_Life_Culture_and_Mentality_of_ Medieval_People. Accessed on June 23, 2019.

Lawson, Erich. 2018. "8 Effective Ways to Reduce Manufacturing Waste." Fishbowl. https://www.fish bowlinventory.com/blog/2018/01/31/8-effective-ways-to-reduce-manufacturing-waste/. Accessed on July 3, 2019.

LeBlanc, Rick. 2019. "Integrated Solid Waste Management (ISWM)—An Overview." Small Business. https://www .thebalancesmb.com/integrated-solid-waste-management-iswm-an-overview-2878106. Accessed on July 2, 2019.

Lehne, Mark, and Ana Tavares. 2010. "Walls, Ways and Stratigraphy: Signs of Social Control in an Urban Footprint at Giza." In Manfred Bietak, Ernst Czerny, and Irene

Forstner-Müller, eds. *Cities and Urbanism in Ancient Egypt*. Vienna: Österreichische Akademie der Wissenschaften. http://www.gizapyramids.org/static/pdf%20library/lehner_urbanism.pdf. Accessed on June 16, 2019.

Loewen, Chris. 2018. "Gehenna: The History, Development and Usage of a Common Image for Hell." Rethinking Hell. http://rethinkinghell.com/2018/01/23/gehenna-the-history-development-and-usage-of-a-common-image-for-hell/. Accessed on June 21, 2019.

Lofrano, Giusy, and Jeanette Brown. 2010. "Wastewater Management through the Ages: A History of Mankind." *The Science of the Total Environment*. 408(22): 5254. https://www.researchgate.net/publication/46148626_Wastewater_Management_through_the_Ages_A_History_of_Mankind. Accessed on June 21, 2019.

Lu, Shyi-Min. 2018. "Resource Recycling and Waste-to-Energy: The Cornerstones of Circular Economy." *Invention Journal of Research Technology in Engineering & Management*. 2(8): 4–22.

"Materials Generated in the Municipal Waste Stream, 1960 to 2015." 2018. *Advancing Sustainable Materials Management: 2015 Tables and Figures*. Washington, DC: Environmental Protection Agency. https://www.epa.gov/sites/production/files/2018-07/documents/smm_2015_tables_and_figures_07252018_fnl_508_0.pdf. Accessed on July 3, 2019.

"Materials Recycled and Composted in the Municipal Waste Stream, 1960 to 2015." 2018. *Advancing Sustainable Materials Management: 2015 Tables and Figures*. Washington, DC: Environmental Protection Agency. https://www.epa.gov/sites/production/files/2018-07/documents/smm_2015_tables_and_figures_07252018_fnl_508_0.pdf. Accessed on July 4, 2019.

McMahon, Augusta. 2016. "Trash and Toilets in Mesopotamia: Sanitation and Early Urbanism." *The Ancient Near East*

Today 4(4). http://www.asor.org/anetoday/2016/04/trash-and-toilets-in-mesopotamia-sanitation-and-early-urbanism/. Accessed on June 15, 2019.

McMahon, Augusta. 2015. "Waste Management in Early Urban Southern Mesopotamia." In Piers D. Mitchell, ed. *Sanitation, Latrines and Intestinal Parasites in Past Populations*, 19–39. New York: Routledge.

McMillan, Sheryl. 2019. *Water Pollution: Types, Causes and Management Strategies*. New York: Syrawood Publishing House.

"Medical Waste Tracking Act of 1988." 2016. Environmental Protection Agency. https://archive.epa.gov/epawaste/nonhaz/industrial/medical/web/html/tracking.html. Accessed on June 27, 2019.

"Medicine and Health in the Middle Ages." 2019. Dose Spot. https://www.dosespot.com/medicine-and-health-in-the-middle-ages. Accessed on June 23, 2019.

Medina, Martin. 2007. *The World's Scavengers: Salvaging for Sustainable Consumption and Production*. Lanham, MD: AltaMira Press.

Melosi, Martin V. 2002. "The Fresno Sanitary Landfill in an American Cultural Context." *Public Historian*. 24(3): 17–35. https://pdfs.semanticscholar.org/643a/1a3346b02625d6b313f8f336f4b49673f0ca.pdf. Accessed on June 25, 2019.

Melosi, Martin V. 2005. *Garbage in the Cities: Refuse, Reform, and the Environment*. Pittsburgh: University of Pittsburgh Press.

"Mercury-Containing and Rechargeable Battery Management Act." 1996. Public Law 104–142. https://www.epa.gov/sites/production/files/2016-03/documents/p1104.pdf. Accessed on June 27, 2019.

"Michigan Bottle Deposit Law. Frequently Asked Questions." 2019. https://www.michigan.gov/documents/deq/

dnre-whmd-sw-mibottledepositlawFAQ_318782_7.pdf. Accessed on July 5, 2019.

"The Middle Ages 'Roots.'" 2019. The History of Sanitary Sewers. http://www.sewerhistory.org/time-lines/tracking-down-the-roots-of-our-sanitary-sewers/part-2-the-middle-ages/. Accessed on June 21, 2019.

"Municipal Solid Waste Generation, Recycling, and Disposal in the United States: Facts and Figures for 2012." 2012. Environmental Protection Agency. https://www.epa.gov/sites/production/files/2015-09/documents/2012_msw_fs.pdf. Accessed on June 25, 2019.

"National Overview: Facts and Figures on Materials, Wastes and Recycling." 2018. Environmental Protection Agency. https://www.epa.gov/facts-and-figures-about-materials-waste-and-recycling/national-overview-facts-and-figures-materials#Generation. Accessed on June 28, 2019.

Newitz, Annalee. 2017. "Plumbing Discovery Reveals the Rise and Fall of the Roman Empire." Ars Technica. https://arstechnica.com/science/2017/08/plumbing-discovery-reveals-the-rise-and-fall-of-the-roman-empire/. Accessed on June 30, 2019.

"The Past, Present, and Future of Solid Waste Disposal." 2017. Sustaining Our World. https://sustainingourworld.com/2011/09/22/the-past-present-and-future-of-solid-waste-disposal/. Accessed on June 20, 2019.

Platt, Brenda, Nora Goldstein, and Craig Coker. 2014. Institute for Local Self-Reliance. https://ilsr.org/wp-content/uploads/2014/07/state-of-composting-in-us.pdf. Accessed on July 4, 2019.

"Pollution Prevention." 2020. Environmental Protection Agency. https://www.epa.gov/p2/pollution-prevention-law-and-policies. Accessed on March 30, 2020.

"Public Health Act, 1875." 1875. http://www.legislation.gov.uk/ukpga/1875/55/pdfs/ukpga_18750055_en.pdf. Accessed on June 24, 2019.

"Report to Congress: Disposal of Hazardous Wastes." 1974. Environmental Protection Agency. https://nepis.epa.gov/Exe/ZyPDF.cgi/10003IEJ.PDF?Dockey=10003IEJ.PDF. Accessed on June 26, 2019.

"Resource Conservation and Recovery Act (RCRA) Compliance Monitoring." 2019. Environmental Protection Agency. https://www.epa.gov/compliance/resource-conservation-and-recovery-act-rcra-compliance-monitoring. Accessed on June 26, 2019.

"Risks Associated with Feeding Raw or Improperly Cooked Food Waste to Swine." 2018. California Department of Food and Agriculture. https://www.cdfa.ca.gov/ahfss/Animal_Health/pdfs/GarbageFeedingFactsheet.pdf. Accessed on July 2, 2019.

Rosbe, William L., and Robert L. Gulley. 1984. "The Hazardous and Solid Waste Amendments of 1984: A Dramatic Overhaul of the Way America Manages Its Hazardous Wastes." *Environmental Law Reporter*. https://elr.info/sites/default/files/articles/14.10458.htm. Accessed on June 27, 2019.

Rosen, George. 1958. *A History of Public Health*. New York: MD Publications.

Routt, David. n.d. "The Economic Impact of the Black Death." EH.net. https://eh.net/encyclopedia/the-economic-impact-of-the-black-death/. Accessed on June 23, 2019.

Rusong, Wang. 2002. "System Consideration of Eco-sanitation in China." EcoSanRes. http://www.ecosanres.org/pdf_files/Nanning_PDFs/Eng/Wang%20Rusong%2001_C23x.pdf. Accessed on June 15, 2019.

"Sanitary Landfill." 2019. UNTERM. https://unterm.un.org/UNTERM/Display/record/UNHQ/sanitary_landfill/3A66E75DE35E0F3A85257065006664E7. Accessed on June 25, 2019.

"Secure Landfill Disposal." n.d. Environmental Technology Council. http://etc.org/advanced-technologies/secure-landfill-disposal/. Accessed on July 6, 2019.

"The Shattuck Report." 1850. Historic Public Health Books. https://biotech.law.lsu.edu/cphl/history/books/sr/. Accessed on June 24, 2019.

Shaw, Ian. 2012. "The Archaeology of Refuse Disposal in New Kingdom Egypt: Patterns of Production and Consumption at El-Amarna." *Talanta*. 44: 315–333. https://www.academia.edu/5060255/_The_archaeology_of_refuse_disposal_in_New_Kingdom_Egypt_patterns_of_production_and_consumption_at_el-Amarna_Talanta_Proceedings_of_the_Dutch_Archaeological_and_Historical_Society_2012_vol._XLIV_Special_Research_Issue_pp.315–333. Accessed on June 16, 2019.

Smedley, Tim. 2019. *Clearing the Air: The Beginning and the End of Air Pollution*. London: Bloomsbury Sigma.

Sridevi, V., et al. 2012. "A Review on *[sic]* Integrated Solid Waste Management." *International Journal of Engineering Science & Advanced Technology*. 2(5): 1491–1499. https://pdfs.semanticscholar.org/3475/5d7c9d2df44e803fa7a4799 2254f33f45759.pdf. Accessed on July 2, 2019.

Stott, Romie. 2016. "Oxyrhynchus, Ancient Egypt's Most Literate Trash Heap." *Atlas Obscura*. https://www.atlas obscura.com/articles/oxyrhynchus-ancient-egypts-most-literate-trash-heap. Accessed on June 16, 2019.

Stringfellow, Thomas. 2014. "An Independent Engineering Evaluation of Waste-to-Energy Technologies." Renewable Energy World. https://www.renewableenergyworld.com/articles/2014/01/an-independent-engineering-evaluation-of-waste-to-energy-technologies.html. Accessed on July 5, 2019.

"Sustainable Materials Management: The Road Ahead." 2009. Environmental Protection Agency. https://www.epa.gov/sites/production/files/2015-09/documents/vision2.pdf. Accessed on July 2, 2019.

"Then and Now: The Evolution of Recycling." 2019. *American Lifestyle*. https://americanlifestylemag.com/

culture/green-living/then-and-now-the-evolution-of-recycling/. Accessed on June 25, 2019.

Tiwary, Abhishek, Ian Williams, and Jeremy Colls. 2019. *Air Pollution Measurement, Modelling and Mitigation*, 4th ed. Boca Raton, FL: CRC Press.

Tiwary, Sachin Kr., and Shubham Saurabh. 2018. "Archaeological Evidences of Toilet System in Ancient India." Heritage: *Journal of Multidisciplinary Studies in Archaeology*. 6: 764–781. http://www.heritageuniversityofkerala.com/JournalPDF/Volume6/36.pdf. Accessed on June 30, 2019.

"25 Years of RCRA: Building on Our Past to Protect Our Future." 2002. Environmental Protection Agency. https://archive.epa.gov/epawaste/inforesources/web/pdf/k02027.pdf. Accessed on June 26, 2019.

"2.1 Billion People Lack Safe Drinking Water at Home, More Than Twice as Many Lack Safe Sanitation." 2017. World Health Organization. https://www.who.int/news-room/detail/12-07-2017-2-1-billion-people-lack-safe-drinking-water-at-home-more-than-twice-as-many-lack-safe-sanitation. Accessed on June 30, 2019.

"UN Warns Poor Waste Management Threatens Human Settlement Globally." 2019. China.org.cn. http://www.china.org.cn/world/Off_the_Wire/2019-05/28/content_74831887.htm. Accessed on June 15, 2019.

Wakeling, Bonnie. 2019. *Water Pollution: Effects and Mitigation Strategies*. Berlin: Callisto Publishers.

"Was 'Gehenna' a Smoldering Garbage Dump?" 2011. The Baker Deep End Blog. https://bbhchurchconnection.wordpress.com/2011/04/06/was-gehenna-a-smoldering-garbage-dump/. Accessed on June 16, 2019.

"Waste Disposal and Recycling. Reduce, Reuse, Recycle." 2019. SlidePlayer. https://slideplayer.com/slide/5954905/. Accessed on July 4, 2019.

Waste Incineration and Public Health. 2000. National Research Council. Committee on Health Effects of Waste

Incineration. National Research Council. Board on Environmental Studies and Toxicology. National Research Council. Commission on Life Sciences. Washington, DC: National Academy Press.

Whitten, Harvey. 2009. "The Rittenhouse Paper Mill Circa 1690. . . the Beginning in America." Paper Age. http://www.paperage.com/issues/nov_dec2009/11_2009of_interest_rittenhouse.pdf. Accessed on June 25, 2019.

Wright, Lawrence. 1960. *Clean and Decent: The Fascinating History of the Bathroom & the Water Closet and of Sundry Habits, Fashions & Accessories of the Toilet Principally in Great Britain, France, & America*. London: Routledge & Kegan Paul. Text available through the Hathi Trust Digital Library. https://babel.hathitrust.org/cgi/pt?id=miun.agr4207.0001.001&view=1up&seq=1. Accessed on June 21, 2019.

Zapotoczny, Walter S. 2006. "The Political and Social Consequences of the Black Death, 1348–1351." http://www.wzaponline.com/yahoo_site_admin/assets/docs/BlackDeath.292130639.pdf. Accessed on June 23, 2019.

Zarin, Daniel J. 1987. "Searching for Pennies in Piles of Trash: Municipal Refuse Utilization in the United States, 1870–1930." *Environmental Review*. 11(3): 207–222.

"Zero Waste Definition." 2018. Zero Waste International Alliance. http://zwia.org/zero-waste-definition/. Accessed on July 4, 2019.

"Zero Waste Versus Recycling: What's the Difference?" n.d. Sustainable Jungle. https://www.sustainablejungle.com/sustainable-living/zero-waste-future/. Accessed on July 4, 2019.

Ziegler, Philip. 2017. *The Black Death*. Reprint of the 1969 edition. New York: Harper Perennial.

Zimring, Carl A., and William L. Rathje, eds. 2012. *Encyclopedia of Consumption and Waste: The Social Science of Garbage*. Thousand Oaks, CA: SAGE Publications.

Running Out of Room

One of the most serious crises facing the solid waste management system in the United States is loss of landfill capacity. In 2015, the most recent year for which data are available, the nation generated 262.43 million tons of solid wastes. Just over half (137.7 million tons) of that trash was deposited in landfills. That option is quickly diminishing, however. Recent studies have showed that the net capacity of the nation's landfills has begun to decrease and will continue to do so in the future. In one such study the net capacity of the nation's landfill was found to have *increased* by about 105 million tons in 2016 and then begun to *decrease* by more than 150 million tons a year later. Losses in net capacity of landfills are expected to reach more than 308 million tons in 2021, a net total decrease in landfill capacity of 1.086 billion tons between 2016 and 2021. And there appears to be little hope that this trend will change in the near future (Thompson 2018). So the basic question is this: What is the nation do to do with the enormous amount of solid waste generated each year?

Several options for dealing with this issue are available. Greater use of recycling, compositing, and incineration are examples of such options. Historically, however, another option has been the exportation of solid wastes to other countries around the world.

A digger sits atop a large garbage pile in a landfill. (Vchalup/Dreamstime.com)

Exportation of Hazardous Wastes

On December 31, 1987, a freighter called the *Khian Sea* arrived in Haiti and obtained a permit from the government to unload a portion of its cargo, described to officials as "fertilizer." The material was actually ash produced at a solid waste incinerator in Philadelphia. Between January 20 and January 31, the ship off-loaded 4,000 tons of the ash, only to be interrupted by an order from the Haitian government to reload its cargo and to return to sea. The *Khian Sea* did return to sea, but without reloading the ash it had already left on Gonaives beach. The majority of that ash was to remain on the beach until it was removed in 2002 by the U.S. waste management firm of Waste Management, Inc.

The Haitian episode was only one event in a longer two-year history of the *Khian Sea*'s travels. That history began on September 5, 1986, when the ship left Philadelphia with a 14,000-ton load of incinerator ash from a municipal incinerator in the city of Philadelphia's Roxborough neighborhood. The incinerator, like most municipal incinerators, was used to burn household wastes consisting of everything from newspapers and garbage to cigarettes and dead batteries. Lacking space within city limits to dispose of the incinerator ash, Philadelphia contracted with outlying areas (primarily in New Jersey) to accept its wastes. Although the materials burned in the Roxborough incinerator were relatively benign products of home and municipal use, the ashes that resulted were not. They contained by-products, such as dioxins, furans, cadmium, and lead, considered to be toxic, teratogenic, carcinogenic, or otherwise hazardous to human health.

In 1984, the state of New Jersey informed authorities in Philadelphia that it would no longer accept the city's incinerator ash for disposal. Instead, the city contracted with the firm of Paolino and Sons, Amalgamated Shipping and Coastal Carrier, to dispose of the ash. Paolino's original intent was to dispose of the wastes on a man-made island in the Bahamas, except that the Bahamian government denied the company a permit to unload its cargo on the island. Over the next year, the *Khian Sea* roamed the Caribbean looking for a place to dispose of the ash. Between September 1986 and August 1987, it was turned away from Bermuda, the Dominican Republic, Honduras, Guinea-Bissau, and the Netherlands Antilles. Efforts were then made to convince the Panamanian government to accept the ash for use in road-building in Panama, an offer that was refused.

After leaving Haiti, the *Khian Sea* set off once more to find a dumping ground for the remaining ash in its hold. Over the next six months, it was refused permission to unload its cargo in Morocco, Senegal, Sri Lanka, and Yugoslavia, where it put in for repairs in August 1988. When it was escorted from the country by the Yugoslav navy, it had a new name, the *Felicia*, and a new country of registry, Honduras. Before long, its name was changed once more, this time to the *Pelecano*. On September 28, 1988, it passed through the Suez Canal, listing Singapore as its destination.

By the time the *Khian Sea/Felicia/Pelecano* reached Singapore in November 1988, the troublesome ash cargo had disappeared. The captain later admitted that the ash had been dumped into the Indian and Atlantic Oceans on its way from North America to the Pacific. In 1993, William P. Reilly and John Patrick Dowd, officers of Coastal Carriers Inc., which operated the *Khian Sea*, were convicted of perjury in the United States and sentenced to jail. No convictions were ever obtained for the dumping of ash in Haiti or in the open seas.

In 1997, a coalition of the Greenpeace environmental organization and Haitian environmental groups joined to form the Return to Sender project, aimed at having Philadelphia's hazardous ash removed from Gonaives beach. The project was eventually able to obtain funds from the city of Philadelphia to hire Waste Management, Inc. (WMI) to complete the task. In April 2000, WMI began the task and eventually removed the remaining 2,500 tons of waste to a docking point in Florida's St. Lucie Canal. After resting at that location for two years, the ash was returned to its final burying ground in Franklin County, Pennsylvania, by WMI in June 2002 ("Philadelphia Ash Dumping Chronology" 2008; Schwartz 2000).

The *Khian Sea* experience was only one, although certainly the most dramatic, example of developed nations looking for places on the shores of developing nations to dump their hazardous wastes. A campaign to end this practice eventually resulted in the adoption of an international treaty, the Basel Convention on the Control of Transboundary Movements of Hazardous Wastes and Their Disposal, also known simply as the Basel Convention. The Basel Convention was signed in Basel, Switzerland, on March 22, 1989. It was an agreement among 103 nations establishing certain requirements for the shipment of hazardous wastes between nations. In particular, it required that a nation desiring to ship hazardous wastes to a second nation notify that nation of its intent and that the receiving nation acknowledge its willingness and consent to receive the shipment (Jaffe 1995).

As of late 2019, 186 nations and the European Union have ratified that treaty. The only nations that have not yet done so are Haiti and the United States. A variety of arguments have been presented for America's reluctance to join the Basel Treaty. Some are based on technical issues involving the precise definition of "hazardous materials" used in the treaty and the U.S. RCRA. Others have a more economic focus, arguing that developed nations have the right to offer good trading opportunities with developing nations to make money from accepting harmful products about which they care less than do developed nations. (For one of the most famous examples of this position, see "The Memo" 1991; also see Schmidt 1999.)

The U.S. government currently believes that it has adequate rules and regulations in place for the shipment of hazardous wastes from this country to other nations around the world. Some of those provisions are enshrined itself in the RCRA ("Information for Exporters of Resource Conservation and Recovery Act (RCRA) Hazardous Waste" 2019), while others are part of agreements between the United States and one or more other countries (currently Canada, Mexico, Costa Rica, Malaysia, and the Philippines (International Agreements on Transboundary Shipments of Hazardous Waste 2019). As a consequence, the Environmental Protection Agency regards the export of hazardous wastes as essentially a "nonissue" for our government.

The transport of hazardous wastes from the United States to some other country is no longer a viable alternative. Over the past few decades, therefore, researchers have developed other mechanisms for disposing of hazardous waste. One method makes use of chemical reactions that convert the chemicals responsible for a material's toxicity to more benign products. The products of this reaction can then be treated with methods like those of nonhazardous wastes. Some biological systems are also available for achieving the same result.

Far more common than chemical or biological methods are physical procedures. In one such process, hazardous materials are encased in concrete containers and buried in secure landfills. Safety systems associated with the landfills monitor any changes in hazardous materials that may require further treatment. By far the most common method used for hazardous materials storage, however, is deposit in deep wells that are no longer used for other purposes. Again, the hazardous materials must be securely encased before their deposit in a deep well. According to recent data, deep well injection was used for 90–96 percent of all hazardous waste disposal from 2001 to 2015 ("Wastes" 2018; see "ROE Indicators," available at https://cfpub.epa.gov/roe/indicator.cfm?i=54; Nathanson 2019; the EPA has stringent requirements for such systems, however; see "Underground Injection Control (UIC)" 2016).

Exportation of Plastic Wastes

Hazardous wastes are by no means the only form of solid wastes for which exportation has seemed to be a solution. Over the last few decades, shipment of *plastic wastes* from the

United States to other countries has become an issue of special importance. Recycling and other forms of disposal for plastics has been the least popular of all forms of solid waste disposal. In 2015, for example, less than 10 percent of all plastic wastes were recycled, at a time when comparable numbers for paper and yard trimmings were 67 and 61 percent, respectively ("National Overview: Facts and Figures on Materials, Wastes and Recycling" 2018).

The possibility of recycling plastic waste first became popular in China in the mid-1990s. In 1995, entrepreneur Zhang Yin established a company, Nine Dragons, for recycling paper, an endeavor that was hugely successful, ultimately making Zhang the first female billionaire in China. She then decided to expand her business to include plastic recycling, an operation that also became very successful; countries around the world, including the United States, were looking for ways to dispose of the large amounts of plastic wastes they were accumulating. By 2016, Zhang and her Chinese colleagues were importing more than 700,000 metric tons of plastic wastes a year, a large, but paltry, sum compared to imports from Hong Kong (1.78 million metric tons) and Japan (842,000 metric tons) (Ritchie and Roser 2018). For the United States (and many other countries), China's willingness to accept huge amounts of plastic waste was a godsend.

But, then, Chinese attitudes about plastic wastes began to change. By 2016, the government had begun to question the wisdom of taking in trash that was often contaminated by large amounts of nonplastic materials, increasing the costs of recycling the plastic alone. It also began to take the position that China was concerned about being overwhelmed by waste plastics, as had previously been the case in most other developed nations of the world. As a result, in 2017, it announced that it would no longer accept plastic wastes from other parts of the world. Within a year, the amount of that material accepted by China dropped to less than 1 percent of the quantity take in a year previously. The successful use of China as the world's plastic garbage dump had come to an end. It continued to produce about 20 percent of the world's plastic wastes from its own industries, however (Parker 2018; Wang et al. 2019).

What was to become of the 275 million metric tons of waste plastic produced worldwide every year? One answer was to find other countries that *would* accept plastic wastes. Today, that list is fairly small and includes primarily South East Asian nations such as India, Malaysia, South Korea, Vietnam, and Thailand. In 2018, the largest quantity of waste plastics went to Malaysia (187.7 million pounds), followed by Hong Kong (122 million pounds), Thailand (91.5 million pounds), and Vietnam (65.4 million pounds). At the same time, Chinese imports of waste plastics had dropped to 50 million pounds (Staub 2018). The utility of that means of plastic waste disposal has begun to dry up very quickly, however, as some of the countries on the above list have now decided to discontinue accepting at least some forms of waste plastic. Malaysia has also begun sending back plastic wastes to the countries from which they came, while Thailand and Vietnam have decided to ban all future imports of waste plastics by 2021 and 2025, respectively (Harris, Latiff, and Birsel 2019; Niranjan 2019; Zhou 2019).

Plastic Wastes

In 1997, Captain Charles J. Moore was returning to the United States from his participation in the Los Angeles to Hawaii Transpac sailing race. He decided to take a shortcut across the mid–Pacific Ocean that was somewhat less well known to sailors. On this voyage he came across a great expanse of trash floating on the water. The "great expanse" eventually turned out to be more than twice the size of the state of Texas, covering an area of more than 600,000 square miles. The area has since come to be known as the Great Pacific Garbage Patch (GPGP) (Moore 2014).

The GPGP actually consists of two parts, the Eastern and Western Garbage Patches, connected by an ocean current about 6,000 miles in length. The region has been subjected to extensive research since its discovery. Scholars now believe that the garbage patch is made up of about 1.8 trillion pieces of plastic weighing more than 87,000 tons. Much of the plastic trash has come from land-based disposal, although the majority appears to be fishing-related items, such as nets, abandoned at sea ("Great Pacific Garbage Patch" 2019). The effect on sea life is devastating, with an estimated 1 million dead seabirds and 100,000 sea mammals lost every year to materials found in the GPGP (Grant 2009).

The garbage patch illustrates another significant feature of the world's waste management problem: plastic wastes. Plastic wastes pose a somewhat different problem than do other forms of solid wastes. They are, in most cases, very stable compounds that remain in a landfill for dozens or hundreds of years. That stability also means that they tend to pose technical problems for recycling processes and are not good candidates for composting. If incinerated, they produce a host of toxic and otherwise harmful by-products—such as bisphenol A; phthalates; dioxins; furans; polychlorinated biphenyls (PCBs); and heavy metals such as cobalt, copper, chromium, cadmium, lead, and mercury (Verma et al. 2016). Their persistence in the oceans, as in the GPGP, is only one example of their long lives in the environment. Efforts to dispose of plastic wastes have, therefore, often involved simply "throwing them away" in ways that they are washed into rivers and then, eventually, into the oceans. The basic philosophy behind such practices is usually "out of sight, out of mind." If plastic wastes are not readily visible, they cannot really be much of a problem. GPGP and other discoveries of plastic wastes in the oceans, however, put an end to that philosophy (Abbing 2019; "Fighting for Trash-Free Seas" 2019; Marine Plastic Pollution 2019).

Given all the problems listed above, the one remaining option for controlling plastic pollution appears to be at the level of generation: reducing the amount of plastic and plastic products that are made each year may be the best single way to manage plastic wastes. And within that framework, the greatest attention has been paid to so-called *single-use* products. These products are items that are intended to be used only once and then discarded. Examples of such products include cigarette butts (the most common of all single-use products), plastic drinking bottles, plastic bottle caps, food wrappers, plastic grocery bags, plastic lids, straws and stirrers, other types of plastic bags, and foam takeaway containers ("Single-Use Plastics: A Roadmap for Sustainability 2018"; this may be the best single existing resource on single-use plastics).

Programs to eliminate or reduce the use of single-use plastics are slowly gaining momentum in the United States and in many other countries. Some of the most important of these steps include the following.

- More than 400 cities in the United States alone have banned the use of plastic bags (Nance 2018).
- One hundred twenty-seven countries worldwide have adopted some form of restriction on the use of plastic bags ("Legal Limits on Single-Use Plastics and Microplastics: A Global Review of National Laws and Regulations" 2018).
- Twenty-seven countries have banned single-use plastics in other items, such as straws, cups, plates, or packaging ("Legal Limits on Single-Use Plastics and Microplastics: A Global Review of National Laws and Regulations" 2018).
- The Aquarium Conservation Partnership (ACP) announces November 2018 to be "No Straw November," a project that encourages more than 500 businesses to abandon the use of single-use plastic straws. ACP is a consortium of twenty-two museums working for the protection of marine mammals ("Newport Aquarium Joins National Push to Take #FirstStep to Cut Plastic Pollution" 2018).
- The Ellen MacArthur Foundation announced the creation of the New Plastics Economy Global Commitment, a program designed "to address plastic waste and pollution at its source." Among the more than 250 organizations who have signed on to the project are the Coca-Cola Company; World Wildlife Foundation; Walmart, Inc.; European Investment Bank;

the governments of Chile, France, New Zealand, and United Kingdom; the cities of Austin (Texas) and Ljubljana (Slovenia); Bangor University; and the Solid Waste Association of North America ("'A Line in the Sand'—Ellen MacArthur Foundation Launches New Plastics Economy Global Commitment to Eliminate Plastic Waste at Source" 2018).

- One of the world's most prodigious users of single-use plastics, the airline industry, has begun to reduce or eliminate the use of such products. Among the airlines that have joined this effort thus far include Alaska, American, Delta, Virgin, United, Qantas, British Airways, Emirates, SAS, Iberia, Ryanair, and Hi Fly (Chua 2019).
- Corona beer announced the development of interlocking beer cans to replace the traditional single-use six-pack rings. It is only one of several companies developing alternatives to plastic rings (Pomranz 2018a, 2018b). (An excellent list of ideas for reducing use of plastics in one's everyday life can be found at "Tips to Use Less Plastic" 2018.)

Nuclear Wastes

On March 3, 2010, the U.S. Department of Energy (DOE) sent a notice to the Atomic Safety and Licensing Board of the U.S. Nuclear Regulatory Commission saying that it was withdrawing its application for the use of a site at Yucca Mountain, Nevada, for the storage of nuclear wastes (U.S. Department of Energy's Motion to Withdraw 2010). In its application, the DOE said that the site was "not a workable option for long-term disposition" of spent nuclear fuel. Thus ended a nearly half-century search for a method to dispose of nuclear wastes. The effort went back to the mid-1950s, when it had become apparent that some mechanism was needed for safely storing nuclear materials with lives in the tens and hundreds of thousands of years.

Nuclear (or radioactive) wastes are of two general types: low level and high level. (Other categories—intermediate level, mining tailings, and transuranic wastes—are also placed in a category of their own.) Low-level wastes are by far the most common type of nuclear waste. They are radioactive materials left over from procedures conducted in medical and dental offices, hospitals, educational and research institutions, and private or government laboratories. Some examples of low-level wastes are shoe covers and clothing; cleaning rags, mops, filters, and reactor water treatment residues; equipment and tools; medical tubes, swabs, and hypodermic syringes; and carcasses and tissues from laboratory animals (as quoted in "Low-level Radioactive Waste (LLW)" 2019).

High-level nuclear wastes are the materials left over from operations that take place in nuclear power plants. The fuel rods in which nuclear reactions occur in such plants must be replaced at some point after they have been in use. They still contain signification amounts of uranium, plutonium, and other radioactive materials at that point and must be stored out of contact with humans and the natural environment.

Low-level wastes are so named because their half-lives are of the order of a few dozen years at most. *Half-life* is the time it takes for half of a sample of a radioactive material to disintegrate. If one has a material with a half-life of five days, for example, only half the amount of that material will remain after five days, a quarter after ten days, an eighth after fifteen days, a sixteenth after twenty days, and so on. In general, therefore, low-level radioactive materials pose a threat to human life and the environment for anywhere from a few days to a few dozen years. By contrast, high-level wastes have much longer half-lives, ranging often in the millions of years. These wastes, therefore, are dangerous for tens of millions of years or more.

Because of the relatively short-term, low-level radiation they produce, finding storage sites for these materials is relatively easy. In the United States today, four facilities are licensed to receive and store low-level wastes: Andrews County, Texas; Barnwell, South Carolina; Clive, Utah; and Richland, Washington. The same cannot be said for high-level wastes. Such materials must be interred for very long periods of time during which they are emitting very high levels of radiation. In case of an accident, either these materials or the radiation they produce, or both, may be released to the surrounding environment posing a threat not unlike that of a nuclear plant meltdown and, perhaps, even an atomic weapon detonation.

The U.S. government has been aware of the issues created by the need to dispose of high-level nuclear wastes for more than a half century. As early as 1956, a committee of the National Academy of Sciences recommended that the safest repository for high-level wastes would be an underground site, such as an abandoned mine or natural cave. For the next twenty-five years, experts debated the more precise characteristics of a burial site and began to look for possible locations for a repository. In 1982, the U.S. Congress passed the Nuclear Waste Policy Act of 1982, which described in detail the steps the Department of Energy was to take in order to find one or more disposal sites.

Over the next five years, officials listed first nine and then five possible sites for a repository. Then, in 1987, a joint committee of the Congress named a single location, Yucca Mountain, in Nevada to be the high-level waste depository, The next two decades were taken up with feasibility studies for the Yucca Mountain site along with often vigorous debates between the federal government on the one hand and the state of Nevada and Nye County (where the site is located) and counties adjacent to the site on the other. While the scientific research continued to produce somewhat controversial results, the intergovernmental debates finally became too great for planners. In 2010, the DOE issued the memorandum mentioned above, discontinuing plans for using Yucca Mountain as a high-level nuclear repository. (For a chronology of events and details of the search process, see "Timeline: 1954–2016" 2019; Swift 2015).

So, what is the status of high-level nuclear wastes in the United States today? Lacking any national facility for storage, these wastes are now being kept on a "temporary" basis at thirty-five locations around the United States. In the majority of these cases, wastes are being stored at the site where they are being produced. Some examples include the Columbia Generating

Station in Washington, the Seabrook Nuclear Reactor, and the West Valley (New York) Demonstration Project ("Disposal of High-Level Nuclear Waste" 2019).

As one review of the current status of this issue states, the problems involved in finding a permanent location for high-level nuclear wastes is "social and political opposition," "not technical issues" ("Disposal of High-Level Nuclear Waste" 2019). Thus, the debate over high-level waste siting continues to shift with a variety of actions by the U.S. Congress, decisions by federal courts, and policy positions by Democratic and Republican administrations. The most recent action on the question, as of late 2019, is a request by President Donald Trump for $120 million for a renewal of licensing for the Yucca Mountain location as a site for storage (Smith 2017; for more information on other developments at the Yucca Mountain site, see Rindels and Sanchez 2019).

Electronic Wastes

Each new technological development in human civilization brings with it new problems and new challenges for human health and the natural environment. Just as this occurred for nuclear science in the early 1950s, so it has also occurred in more recent decades with the invention and expansion of electronic technologies. One can hardly imagine a world today without desktop and laptop computers, televisions, DVD players, mobile phones, iPods, iPads, cameras, fans, ovens, washing machines, game consoles, printers, radios, answering machines, pagers, artificial pacemakers, digital scanners, and the like. But, as with all other types of products, any discussion of electronic devices must also include a consideration of their ultimate fate: Do they end up in landfills, incinerators, recycling systems, or some other waste management program?

One might assume that e-wastes are a matter of relatively little concern in the United States. They make up less than 2 percent of all solid wastes generated in this country every year ("National Overview: Facts and Figures on Materials, Wastes and Recycling" 2018). They do, in fact, represent an issue of considerable concern, however, because they contain several chemicals that pose serious health and environmental problems. In this regard, they represent more than 70 percent of all *hazardous* wastes ("E-Waste Facts We All Need to Know" 2019). They are often situated within the complex structure of an electronic device in relatively small quantities. Some examples of those chemicals and their effects are shown in table 2.1. Breaking down a computer, mobile phone, or other electronic device is a time-consuming job that can expose a person to all these chemicals. However, some other e-waste components, such as gold, silver, and rare earth elements, can make that process worthwhile. Major structural parts of an electronic device—such as glass, plastic, and other metals—may also have considerable value for recycling.

Table 2.1 Hazardous Chemicals Commonly Found in Consumer Electronics

Chemical	Health Effect(s)
Arsenic	Dermal, gastrointestinal, respiratory, neurological, and hepatic disorders; carcinogenic
Brominated flame retardants	Possible deleterious effects on blood, breast milk, and body fat
Chromium	Respiratory effects, such as asthma, chronic bronchitis, and chronic irritation; skin disorders, including dryness, scaling, and swelling; lung, nasal, and sinus cancer; and possible hematological toxicity
Cobalt	Dermatitis; lung problems; disorders of the heart
Copper	Gastrointestinal and digestive problems
Lead	Neurological, renal, cardiovascular, hematological, reproductive, and developmental disorders
Mercury	Brain damage, kidney failure, and fetal complications
Polychlorinated biphenyls (PCBs)	Skin problems; disruption of reproductive functions; neurological problems; developmental disorders; thyroid toxicity; carcinogenic effects; and possible cardiovascular, gastrointestinal, immune, and musculoskeletal abnormalities
Polyvinyl chloride	Respirator disorders; possible low-level carcinogenic effects

Source: "Environmental Health and Medicine Education." Various dates. Agency for Toxic Substance and Diseases Registry. Various links, see https://www.atsdr.cdc.gov/substances/index.asp. Accessed on July 12, 2019.

Since at least the 1970s, the most active electronic waste recycling country in the world has been China (Perkins et al. 2014). At one point, the Chinese were importing more than 70 percent of all the electronic wastes being produced worldwide ("70% of Annual Global E-waste Dumped in China" 2012). In 2000, the Chinese government issued an order prohibiting further importation and processing of e-wastes. That order has, however, been largely ignored, and recycling of e-wastes has remained an important commercial enterprise in some parts of the country. According to one estimate, more than 100,000 people are still engaged in e-waste recycling, largely in Guangdong Province in Southern China ("Garbage in China" 2019). According to one estimate, more than 80 percent of the residents of the city of Guiyu in Guangdong make their living from e-waste recycling (Ni and Zeng 2009; an excellent video on Chinese recycling of e-wastes is available at "China—World's Dumping Ground for Electronic Waste (CNN)" 2013).

So, what does the future hold for the recycling of e-wastes in the United States and worldwide?

One review of the problem has noted that "[i]f ocean plastic pollution was one of the major environmental challenges we finally woke up to in 2018, the ebb and flow of public opinion could and should turn to electronic waste in 2019" (Ryder and Zhao 2019). That assessment is based, on the one hand, simply on the volume of wastes involved. According to the best estimates available for 2016 (the latest year for which data are available), about 44.7 million tons of e-wastes were generated worldwide (Baldé et al. 2017). That number is projected by some sources to reach 120 million tons, or more, by 2050 (Ryder and Zhao 2019). (Estimates and projections for e-wastes are problematic because of the reluctance or inability of many nations to keep accurate records regarding this.)

The fundamental e-waste issue that arises is that only a fraction (about 20 percent) of all such materials are captured for recycling. Almost all of the remaining wastes (76 percent) are simply unaccounted for. This fact explains the contention by many experts that e-wastes are not only a serious problem but also a promising opportunity for the future. Under the proper conditions of recycling (which generally do not exist today), e-wastes can be treated to recover valuable chemicals. As an example, a ton of cell phones contains one hundred times more gold than a ton of gold ore, and the silver recovered from e-waste recycling is three times all the silver obtained from mines worldwide each year ("A New Circular Vision for Electronics: Time for a Global Reboot" 2019). One of the most attractive targets of e-waste recycling is rare earth minerals. The term *rare earth* applies to a group of fifteen chemical elements that are not really so much "rare" as they are difficult to extract and separate from each other.

With the increase in the production of electrical and electronic products, rare earth minerals have come under much greater demand. For most of recent history, well over 90 percent of that demand has been satisfied by China. That pattern has begun to change, however, as China has dramatically reduced its exportation of rare earth minerals to the rest of the world. As a consequence, nations are now turning more aggressively to the recycling of e-wastes to extract the rare earth minerals they contain (Tsamis and Coyne 2015).

The risk and opportunity features of e-wastes have inspired the development of a new approach to the recycling process, an approach known as a *circular economy*. This way of thinking about the production, distribution, use, and recycling of devices may be one of the most revolutionary concepts in waste management today. The philosophy on which a circular economy is based is the opposite of the most popular way of thinking about production today, a way of thinking sometimes described as "take, make, and dispose of." That is, companies that make most electronic appliances today think in terms of manufacturing a product and selling it to consumers with the understanding that, after a period of time, the product will become outdated. At that point, consumers are expected to simply throw the product away.

The circular economy view of production mirrors the way materials in the natural environment are produced, used, and recycled. Those systems are often called natural *cycles*. Most people today know at least something about the carbon cycle, nitrogen cycle, water cycle, and other series of changes that plants and animals go through naturally. In every cycle, a material is produced, developed, and then recycled by some mechanism that ensures it will be born again in its original form.

Applying the concept of a natural cycle to the production of devices means that the first step in that process, the actual production of a device, must take into consideration plans for the nature of its end stage. Rather than being made simply to throw away, a device must be produced with the understanding that useful components can be extracted and reused in the production of new devices similar to the "mother" product. (An excellent explanation of this process is available at "What Is a Circular Economy?" 2017) Although applicable to all fields of waste management, the idea of a circular economy is especially useful to the issue of e-wastes. Because modern electronic devices do contain so many useful components, the original design of such appliances must increasingly include plans for their ultimate recycling (Awasthi et al. 2019; this paper is discussed in some detail at "Solving the E-waste Challenge Requires Global Action" 2019).

Industrial Wastes

Most of the discussion of waste management so far in this book has focused on *municipal* solid wastes—that is, the sum total of all objects and materials that individuals, groups, and companies throw away as a result of their everyday activities. Other types of wastes exist also, one of which is industrial waste. The term *industrial waste* refers to the unused and, generally unusable, by-products of operations that take place in factories, power plants, and similar industrial facilities. Industrial wastes are traditionally subdivided into two major categories:

hazardous and nonhazardous. The definition of *hazardous waste* in the United States is found in the Resource Conservation and Recovery Act of 1976, sections 260 through 273. In general, a substance must ignitable, corrosive, reactive, or toxic in order to be listed as a hazardous waste. Some examples of hazardous wastes are solvents, degreasers, glues, waste inks, cleaning solutions, waste acids, bleaching compounds, compounds of chlorine, unspent explosives, and any substance causing long-term health effects on humans ("Defining Hazardous Waste: Listed, Characteristic and Mixed Radiological Wastes" 2019; the relevant section of the RCRA can be found at "Resource Conservation and Recovery Act (RCRA) Laws and Regulations" 2019). Anything that cannot be defined as hazardous waste is, de facto, nonhazardous waste. It falls into the same general category for waste management procedures as do municipal solid wastes. (A good general reference on industrial wastes, although now somewhat dated, is "A Guide for Industrial Waste Management" 2006.)

How much industrial waste is produced each year? Although that question would appear to be relatively easy to answer, those data are hard to come by. And when they exist, they are subject to all kinds of qualifications. According to some authorities, industrial operations produce a far greater amount of waste products than do municipal sources. One writer has expressed this estimate as a 3–97 ratio, in which industrial wastes account for nearly all (97 percent) of the wastes produced in the United States, compared to 3 percent from municipal sources. The same author explains in detail why this ratio is almost certainly wrong, and, in fact, drastically incorrect (Liboiron 2016). For the purpose of this book, those numbers are even more difficult to interpret since many sources include agricultural and mining wastes in the category of industrial wastes. And those two categories, treated separately here, produce much greater volumes of waste than do factories, power stations, and other industrial facilities. (One commonly cited, but out-of-date reference on this topic is "Managing Industrial Solid Wastes from Manufacturing, Mining, Oil and Gas Production, and Utility Coal Combustion" (1992). Perhaps the most frequently cited number of total industrial wastes is the 1987 estimate of 7.6 billion tons, almost certainly far from correct today. See Zimring and Rathje (2012, 430).

The more critical question about industrial wastes is this: How are they disposed of? In general, nonhazardous waste from industries tend to be handled as are any form of municipal solid wastes. In fact, they often share the same sewage system, landfills, and incineration facilities. As noted above, however, hazardous wastes require special treatment before being deposited in secure landfills, wells, or mines, or being incinerated. (For a more detailed and technical discussion of disposal methods, see Muralikrishna and Manickam 2017.)

Increasingly, planning for hazardous waste disposal is based on the so-called 4 R philosophy. The four *R*s in this type of planning are as follows:

- **R**educe: Design products so that they will contain less useless waste material at the end of their lifetime.
- **R**euse: Rather than throwing something away, plan for using it in other ways.
- **R**ecycle: If neither reduce or reuse is available, or are adequate, find ways to make physical changes in a product so that it can be reused as originally intended or in new ways.
- **R**ecover: Make use of traditional means of waste disposal, such as landfills, composting, and incineration, to get rid of any wastes that cannot be handled by any of the previous three *R*s.

Researchers are always looking for new and efficient methods for dealing with wastes of all kinds. One such approach that involves industrial waste management is the so-called *eco-industrial park*. An eco-industrial park has been defined as "a community of manufacturing and service businesses seeking enhanced environmental and economic performance through collaboration in managing environmental and resource issues, including energy, water, and materials." The concept was first developed in the early 1990s by researchers at Cornell University and Dalhousie University in Nova Scotia (Lowe 2015; also see Eco-Industrial Parks 2006).

As of late 2019, the concept of an eco-industrial park has been made a reality in only a small number of locations: five in India; four in Vietnam; two each in Colombia, Morocco, Peru, and South Africa; and one in China. One is also under development in Thailand. A related type of system, called an *urban industrial synergy*, has also been developed in China, Denmark, Japan, and Puerto Rico (Meylan et al. 2017). Among the most commonly discussed systems is the one in Kalundborg, Denmark. (See Botequilha-Leitão 2012 for the discussion that follows.)

One component of this system is an oil refinery, the waste products of which are transferred to three other facilities: a plasterboard plant, an electric power station, and a sulfuric acid plant. The electric power plant then sends its waste products to other facilities in the region, the plasterboard plant, a cement factory, the city of Kalundborg, local farmers, and a road construction operation. The bio-plant, in turn, provides usable wastes to the region's pig farmers and other agricultural operations. Other eco-industrial facilities are substantially more complex than the Kalundborg example described here. (See, for example, "Eco-Industrial Park Initiative for Sustainable Industrial Zones in Vietnam" 2018.)

Researchers have also explored industrial-municipal plans similar to eco-industrial programs, but not identical to them. These projects are in the planning or construction phases in many parts of the world. One example in the United States is the ReVenture Park in Charlotte, North Carolina. The park will be constructed on a 667-acre piece of land formerly occupied by a textile plant. The site is a former brownfield that was eventually cleaned up and removed from the Superfund list in 2012. It will contain a heat and power plant, center for urban wood recycling, wastewater treatment facility, green technology research laboratory, biofuel plant, ethanol pilot production plant, electric truck assembly facility, anaerobic digestion facility, metal recovery demonstration plant, office space for green and environmental not-for-profit organizations, conservation easement of more than 180 acres, and access to the Carolina Thread Trail ("ReVenture Park—Charlotte's First Eco-Industrial Park" 2014; for a good overview of the current status of eco-industrial facilities in 2018, see "Eco-Industrial Parks & Industrial Ecology" 2018).

Agricultural Wastes

By almost any measure, the largest single human activity in the world is agriculture and dairy farming. Feeding the world's 7.7 billion people requires vast amounts of land and human labor. Because these activities often take place in relatively remote regions, following procedures that have been developed over the ages, often in different forms in different parts of the world, precise data and statistics on these operations are generally difficult to obtain.

In the United States, some information is available directly from research, and other data, from indirect calculations. For example, the U.S. Department of Agriculture reports that 91.7 million acres of land were planted to corn and 45.61 million acres of land to wheat in 2019 ("Corn Planted Acreage Up 3 Percent from 2018. Corn Stocks Down 2 Percent from June Last Year" 2019; "Wheat Data" 2019). Of these and other grain crops, about half ends up as waste. For corn, for example, the leaves, stalks, and corn remaining after plant processing, a material known as *corn stover*, makes up about 45 percent of the total dry weight of corn plants. Similar numbers hold true for wheat straw and other unusable parts of a grain plant (Gould 2007). Although the raw numbers are not as high, waste of fruit and vegetable products prior to reaching the marketplace are also high, typically reaching about 20 percent of all unusable products in the field (Johnson et al. 2018).

Wastes from animal farming come from a variety of sources, such as manure, urine, wasted feedstock, animals that die before harvesting, materials collected from a slaughterhouse, wastewater from animal maintenance, and methane gas. While direct data for the amount of these wastes are generally not available, they can be estimated from other information. For example, the U.S. Department of Agriculture reported that the number of animals slaughtered in May 2019 was 2.9 million cattle, 46,200 calves, 10.3 billion hogs, 209,000 sheep and lambs, 812.6 million chickens, and 19 million turkeys ("Meat Statistics Recent" 2019). The wastes produced by these animals can be estimated by knowing the average amount of manure (as just one type of waste) per animal per day. According to one study, this number is twelve pounds of manure per dairy cattle animal per day, between 0.39 and 2.05 pounds per pig per day (depending on the type of pig), and about 0.05 pounds per chicken per day ("Volumes & Amounts of Manure Produced by Livestock" 2018). The total quantity of manure alone produced daily in the United States per day, then, ranges in the tens of millions of pounds.

Waste management systems are, of course, a traditional and integral part of any agricultural or dairying operation. Plant and animal wastes are a natural and unavoidable part of any such operation. Unlike municipal solid waste management systems, modifying the "generation" part of a system is very difficult, if not impossible. In agriculture and dairying, no company is interested in reducing the quantity or size of plant or animal produced in the first place. But such operations still require systems for managing the very large quantities of waste produced.

Probably the most common management system used in dairying today consists of six related elements: production, collection, transfer, storage, treatment, and utilization. As an example, wastes produced in a barn or milking center (production) are often collected by physical systems (such as troughs or manual labor) and then transferred to one of several possible storage locations, such as a covered storage structure, a waste treatment pond, or composter. They may be treated by natural processes, such as drying or sedimentation, or by chemical or biological processes. The treated wastes may then be returned to the fields as fertilizer, made available for sale, buried in a landfill or incinerated ("Agricultural Waste Management Systems" 2011).

Agricultural Wastes as Biofuels

As with other types of waste management, a major emphasis for agricultural wastes has been to reduce the volume of materials that end up at the end stage of their existence, in some form of disposal process, such as a landfill or an incinerator. An important motivating factor in this direction was the adoption of the Energy Independence and Security Act of 2007 (EISA). That act mandated, among other things, greater use of biomass for the production of alternative forms of energy. The term *biomass* has been defined by the U.S. Department of Energy as "[a]ny organic matter that is available on a renewable or recurring basis, including agricultural crops and trees, wood and wood residues, plants, algae, grasses, animal manure, municipal residues, and other residue materials" ("2016 Billion-Ton Report: Advancing Domestic Resources for a Thriving Bioeconomy" 2016); this is the key reference for much of the following discussion and is indicated by its acronym, BT16. It is the most recent of three reports required by the EISA, the other being "Biomass as Feedstock for a Bioenergy and Bioproducts Industry: The Technical Feasibility of a Billion-Ton Annual Supply" (2005 BTS) and "U.S. Billion-Ton Update: Biomass Supply for a Bioenergy and Bioproducts Industry" (2011 BT2).

One of the major goals of BT16 was to project possible ways in which waste materials from agriculture, forestry, algae, and certain types of energy crops could be used. The term *energy crops* refers to crops that are grown specifically for the purpose of producing energy and not for human or other animal consumption. One popular type of energy crop already in use is

Miscanthus x giganteus (elephant grass; silvergrass) a type of grass grown for use in energy-producing facilities (Heaton et al. 2019). Other important energy crops are *Brassica caranata*, *Camelina sativa*, giant reed (*Arundo donax*), jatropha (*Jatropha curcas*), sorghum (*Sorghum bicolor*), switchgrass *(Panicum virgatum*), and willow (*Salix*) (Panoutsou and Singh 2017).

One of the greatest success stories for this field of research has been the development of ways for making ethanol (ethyl alcohol) from corn grain. Ethanol was first used as a fuel for vehicles as early as 1826. It became popular only in the 1970s, however, largely as a result of the sharply rising price of fossil fuels (petroleum and natural gas). (For a timeline of ethanol use as an automotive fuel, see "Ethanol Timeline" 2011). Since that time, *gasohol*, a mixture of ethanol and conventional gasoline, has continued to be a popular fuel for motor vehicles. In 2015, about 14.1 billion gallons of ethanol were used as a gasoline additive (BT16, table 2.1, p. 19).

An important addition to the list of possible biofuels discussed in BT16, not included in earlier reports, was algae. Algae are plant or plantlike organisms without stems or leaves that grow primarily in water or on damp surfaces. The two forms of algae studied in BT16 were a freshwater organism, *Chlorella sorokiniana*, and a saltwater form, *Nannochloropsis salina*. Estimates for the amount of each species of alga available for energy production was estimated at about 46 million tons for *C. sorokiniana* and 86 million tons for *N salina*. (BT16, table ES.2, p. xxvii). Most studies focus on the use of smaller types of algae, generally known as *microalgae*, compared to larger species, otherwise known as *macroalgae*.

The process of converting algae to energy is relatively simple in concept, if not in practice. The key to this process is the presence of small amounts of oil within each algal cell. Under the proper conditions, that oil can be extracted from cells, sent to a purification facility, and then used as a biofuel (Gramling 2009).

Algal fuels have a number of positive features, compared to other sources of energy. The life cycle of most species is completed in about a week, allowing a much faster harvesting of fuel than is possible with, say, corn for ethanol. This means that the amount of energy produced per acre for algal facilities is perhaps hundreds of times than that of other processes, such as ethanol production. Algal plans can be grown in areas, such as wastewater ponds and garbage dumps, which are otherwise unusable for agriculture or other applications. Carbon dioxide is an essential raw material for algal growth, so algal facilities might be an important feature in any climate-change-reversal program ("Algae Basics" n.d.; for a good overall description of the algal biofuel process and an image of an existing plant, see "All You Need to Know about Algae Biofuel" 2017).

BT16 focuses not only on the use of crops grown specifically for the production of biomass and biofuels but also on the recovery of various types of agricultural wastes for the production of biofuels. The report summarizes possible sources of wastes from agriculture, forestry, and municipal solid wastes and projects the amount of energy that can be produced from each of these possible sources. See table 2.2 for a summary of those projections.

Agricultural Hazardous Wastes

One can easily imagine that agricultural wastes are dirty, smelly, and occupy too much space. But the idea that they can also be hazardous may be somewhat less obvious. Yet, such is certainly the case. Farmers, arborists, and other agriculturists routinely make use of chemicals that have very important benefits for the crops they manage. But those chemicals may also pose threats to human health and the environment. Chemicals that fall into this category include ash from combustion operations; wood maintenance products, such as paints, stains, strippers, cleaners, and adhesives; containers that have been used for hazardous chemicals in the field, in farm buildings, or in residences; vehicle maintenance products, such as oils, grease, leftover fuels, rust-removal chemicals, and solvents; batteries; and, especially, fertilizers and pesticides.

Table 2.2 Summary of Possible Sources of Biofuel from Wastes, Current and Projection*

Waste Type	Current	2017	2022	2030	2040
Animal manures	17.1	18.0	18.5	18.6	18.4
Cotton field residues	3.3	0.9	1.5	1.7	1.7
Cotton gin trash	1.7	1.7	1.9	2.0	2.1
Grain dust and chaff	5.1	0.0	0.0	0.0	0.0
Orchard and vineyard prunings	5.5	5.5	5.6	5.8	6.0
Rice straw	4.3	4.9	5.2	5.4	5.6
Rice hulls	1.2	1.4	1.5	1.5	1.6
Soybean hulls	2.8	0.0	0.0	0.0	0.0
Sugarcane field trash	1.1	1.0	1.1	1.1	1.1
Treatment thinnings, other forestland	0.0	0.0	0.0	0.0	0.0
Mill residue, unused secondary	4.2	4.2	4.2	4.2	4.2
Mill residue, unused primary	0.5	0.5	0.5	0.5	0.5
Urban wood waste**	15/5.1	45/5.3	47/5.3	49/5.3	49/5.3
Biosolids	3.6	3.6	3.8	4.0	4.2
Trap grease	1.0	1.1	1.1	1.2	1.2
Food processing wastes	4.0	4.0	4.0	4.0	4.0
Landfill gas***	–	45	229	229	229
Total					
All agriculture	–	35	35	36	36
All MSW	–	55	55	55	55
All forestry	–	47	47	49	49
All other	–	8.7	8.9	9.2	9.4
Total	–	142	146	149	149

*Projections are provided for three values of a crop, $40, $50, and $60 per dry ton. Numbers included here refer only to the middle of this range, $50 per dry ton.

**Two methods for projection.

***billion cubic feet.

Source: "2016 Billion Ton Report: Advancing Domestic Resources for a Thriving Bioeconomy." 2016. Chapter 5. Waste Resources. Department of Energy. https://www.energy.gov/sites/prod/files/2016/12/f34/2016_billion_ton_report_12.2.16_0.pdf. Accessed on July 15, 2019.

The latter two categories are among the most important because they are used in such large quantities and they contain such potentially dangerous substances. They most commonly escape into the environment when they are washed off the surface of the land and into nearby waterways or they seep into the ground and dissolve in groundwater. In either instance, any hazardous materials contained in waste fertilizer, pesticides, or herbicides becomes part of the natural environment or are ingested by humans.

The most common component of chemical fertilizers is one or more compounds of nitrogen, usually nitrates. Nitrogen compounds are critical to the growth of plants when used in appropriate amounts. But they can be harmful to humans and other animals when present in excessive amounts. In some cases, they may also be converted in the environment to another class of nitrogen compounds known as nitrites. Considerable evidence now exists for the health effects of nitrates and nitrites on humans. Probably the most important is the occurrence of methemoglobinemia, a condition in which the body is unable to transport oxygen in the body at the rate needed. The condition is especially common among babies and young children who are exposed to high levels of nitrates or nitrites in drinking water. Excess nitrates and nitrites in the body have also been associated with developmental issues and the occurrence of certain types of cancer ("Public Health Statement for NITRATE and NITRITE" 2015; Ward 2009).

Nitrogen compounds can also have adverse effects on other animals in the environment. For example, nitrates and nitrites can be converted in aquatic environments to ammonia, NH_3. Ammonia is toxic in even small amounts and has regularly been found to be the causative agent in fish kills of substantial size ("Ammonia" 2019). The effects of dissolved nitrogen in rivers and lakes is perhaps best illustrated in oceanic regions known as *dead zones*. A dead zone is an area in the ocean where essentially no aquatic life is able to survive. One of the best-known dead zones in recent years occurred in the Gulf of Mexico, where the zone covered an estimated 6,000–7,000 square miles. Nitrogen compounds, and those of another major component of fertilizers, phosphorus, are generally thought to be the major cause of dead zones (Bruckner 2018; possibly the best single resource on the environmental and health risks of agricultural wastes is Steinfeld et al. 2006).

People interested in waste management must often ask themselves three important questions about agricultural hazardous wastes. First, how likely is it that a substance used in agricultural operations, such as fertilizers, pesticides, and herbicides, escapes into the environment? Second, what risks do these products pose to the environment in general and to human health in particular? Third, given the answers to these two questions, what are the relative benefits and risks involved in using various agricultural products and what directions does that information provide about dealing with agricultural hazardous wastes?

No definitive answer exists for the first question. The amount of fertilizer or pesticide that is actually retained by a target area depends on a host of factors, including the slope of the ground on which the substance is applied, the tendency of the soil in that area to erode (which is, in and of itself, the result of many other factors, such as the composition of the soil), and the amount and timing of rainfall and irrigation. The application of ten pounds of fertilizer on an acre of ground, then, can result in a wide range of runoff from the area. Researches have attempted to develop mathematical tools to predict fertilizer and pesticide loss for some or all of the possible factors that may be responsible for runoff. (As an example, see Vadas et al. 2009.)

The second question poses many research problems. It is almost always difficult to determine the long-term (and, sometimes, even short-term) effects of any one given chemical on the health of a frog, starling, cow, or human. Most studies in this area rely on probabilities: What is the likelihood that the use of *x* pounds of a compound will harm any given type of organism?

The third question poses less of a challenge for researchers. One can calculate the economic benefit of using a fertilizer, pesticide, or herbicide on a crop fairly easily. The challenge is to compare crop yield *with* use and *without* use. It should be possible then to arrive at least at a general decision, as to the economic benefits of using the substance on a crop. One example of such a study found that the use of five different pesticides resulted in an additional economic benefit to potato growers of anywhere from $52 million to $281 million per year (Guenthner et al. 1999). It is hardly surprising that studies such as these find that the use of fertilizers, pesticides, herbicides, and similar products increases the value of farm crops, sometimes by modest amounts, but often by significant amounts (Popp, Pető, and Nagy 2013).

The problem is that many of the products used to increase agricultural productivity also have harmful effects on the environment and human health. Pesticides are a good example of this truism. Pesticides, herbicides, insecticides, fungicides, and similar products have been developed and are being used, after all, to kill some type of organism. It would be somewhat witless to assume (or hope) that they have no other effects on organisms living in a treated area, or even in a nearby area. Very large amounts of data have now been collected on the health and environmental effects of agricultural products on the environment and human health. Indeed, such data are required by regulatory agencies before they allow the use of such products in the United States.

To take just one example, the most popular herbicide in use today in the United States is glyphosate. About 145,000 tons of the compound were used in the United States in 2019 (Bennett 2019), and nearly 1.8 billion tons have been used since it was first introduced in 1974 (Benbrook 2016). But questions have been raised about the possible environmental and health effects of glyphosate. One recent study, for example, found that the compound appears to produce changes in the microbial communities found in soil, plants, and animal guts. It may also be one of the factors leading to antibiotic resistance. And it has long been thought to be responsible for endocrine disruptions and possible carcinogenic effects in humans (Malkan 2019; Van Bruggen et al. 2018).

The fundamental question that requires an answer then is this: How does one balance the *benefits* of a particular agricultural product against possible *risks* posed to the environment and human health? If we have good information in the case of glyphosate about its positive economic benefits for American agriculture compared to uncertain, but possible, serious effects on bacteria, frogs, starlings, and humans of glyphosate wastes in the environment, what is the best action to take about continued use of this herbicide?

As with many waste management issues, one promising option is to think about the problem at its source: Are there alternative ways for farmers to deal with their weed problems than using large quantities of glyphosate? Or their pest problems with toxic pesticides? For a number of years, more farmers have been moving toward an approach for pest control known as integrated pest management (IPM). Essentially, IPM is based on the adoption of whatever methods and systems that are economically feasible, environmentally sensitive, and socially acceptable. An IPM program has several fundamental elements:

Construction adjustments. Wooden buildings can be built so that they are not in contact with soil, thus reducing the growth of carpenter ants, termites, and other insect pests.

Physical barriers. In many cases, young trees can be encased in linings, screens can be installed on indoor growing sites, and trenching and covers can be installed for row plants.

Pest resistant crops. Some plant varietals are naturally more resistant to pests than are others, and other varietals have been developed to include this property. Such plants can be grown rather than less resistant varietals.

Pheromone trapping. Pheromones are natural insect scents that can be installed in traps, where they attract pests and remove them from a crop setting.

Biological controls. Many instances exist in nature in which organisms that can be classified as pests by human standards have their own natural predators. If those natural predators are released in a field, they will attack and disable or kill their natural prey. For example, a bacterium known as *Bacillus thuringiensis kurstaki* is a natural predator of many varieties of caterpillar, including the agricultural pests cabbage loopers, tomato hornworm, and tent caterpillars. A very important subcategory of biological controls are so-called biological pesticides, or *biopesticides*. These are chemicals that have been extracted from plants and animals that are toxic to one or many types of agricultural pests. The EPA maintains a list of biopesticides that have been studied and registered with the agency. As of late 2019, that list contained 366 items ("Biopesticide Active Ingredients" 2018). The special advantage of biopesticides is that while they are toxic to one or another insect, fungus, or other pest organism, they have no harmful effects on humans, farm animals, or other animals of importance to humans.

Each decision about the disposal of agricultural wastes, then, involves an often complex mixture of scientific, social, political, economic, and other questions. One recent example illustrates this point. The case involves a pesticide known as sulfoxaflor. The compound was approved by the EPA in 2016 for use in insecticides against sap-feeding insects. Approval came along with a recommendation that the compound not be used in regions occupied by plant pollinators, such as the honeybee. This recommendation was based on studies showing that sulfoxaflor was highly toxic to such animals. The EPA recommended that labeling of sulfoxaflor products suggest that users notify beekeepers up to a mile away in time to allow them to provide extra protection for their bees ("EPA Approves Expanded Use of Indiana-Made Insecticide Toxic to Bees" 2019).

On July 10, 2019, the EPA reversed its position on the use of sulfoxaflor. It decided that the economic benefits provided to farmers by use of the insecticide outweighed any possible harm that might come to honeybees. This announcement illustrates how changing political, economic, scientific, and other information can change official policy on the use of agricultural products and the wastes they produce.

Agricultural Waste and Climate Change

Agricultural operations are a major contributor to climate change. Thirty-seven percent of Earth's surface is devoted to agriculture, and agricultural operations are responsible for 52 and 84 percent, respectively, of all anthropogenic methane and nitrous oxide released to the atmosphere. Methane and nitrous oxide are two major greenhouse gases (GHGs) capable of trapping heat reflected from Earth's surface, contributing to a rise in Earth's annual average temperature. Methane and nitrous oxide may not be as well known as carbon dioxide although they are, respectively, twenty-one and 310 times as efficient as carbon dioxide in capturing and retaining heat in the atmosphere ("Greenhouse Gas Emissions from Composting of Agricultural Wastes" 2001).

Agricultural operations contribute to the release of greenhouse gases in several ways. For example, croplands are traditionally managed to obtain the largest yield of a product per acre of land. Environmental factors in management programs may be of little to no concern for the farmer. But new and better methods of managing croplands can change emission patterns for carbon dioxide, methane, and especially, nitrous oxide. For example, crop rotation that involves the use of legume crops may require smaller amounts of chemical fertilizers and, hence, result in reduced amounts of nitrous oxide released to the atmosphere. More precise application of fertilizers to fields may achieve the same results. The use of reduced tillage, or even the elimination of the practice entirely, has been found to help soils retain nutrients that might otherwise be converted to waste products. More efficient management of rice cropland can reduce the amount of GHGs released to the atmosphere. The insertion of "islands" of natural vegetation has been found to increase the efficiency with which lands make use of nutrients (Smith, et al. 2008; there is now a rich collection of literature on the effects of agricultural operations and agriculture wastes on climate change; see, for example,

Gelfand and Robertson 2015; "Global Mitigation of Non-CO_2 Greenhouse Gases: 2010–2030" 2013; Grossi et al. 2019; Hou, Velthof, and Oenema 2015; Johnson et al. 2007; Kleppel 2014; "Livestock's Long Shadow" 2006; Miner 2010; Paustian et al. 2004; Takahashi Young, eds. 2001).

Livestock operations are an issue of special importance because of the very large contribution of enteric fermentation they involve. When a cow, goat, pig, chicken, or other farm animal digests its food, it produces and releases methane into the atmosphere. The amount of methane released varies from species to species, ranging from essentially zero for poultry to 99.6 million tons of methane per year in 2009 for dairy cattle. Comparable values for other animals were 33.2 million tons for beef cattle, 3.6 million tons for horses, 2.1 million tons for swine, 1 million tons for sheep, and 0.3 million tons for goats (Dunkley and Dunkley 2013, table 4, p. 23).

Animal wastes are also an important source of GHG emissions. In this case, the most important contributor of GHG from wastes was the farming of beef cattle, which produced in 2009 24.5 million tons of carbon dioxide and 5.9 million tons of nitrous oxide. Other major contributors were swine waste (19 million tons of carbon dioxide and 2 million tons of nitrous oxide), dairy cattle (2.7 million tons and 7.8 million tons) and poultry (2.7 million tons and 1.6 million tons) (Dunkley and Dunkley 2013, table 4, p. 23).

Researchers have suggested a host of procedures that can be implemented to reduce GHG emissions from animal operations. These include adding additives to animal feed that will moderate the types of chemicals produced in animal waste; modifying feeding schedules to improve food consumption efficiency; development of storage and separation systems for manure; use of best-practice composting methods for wastes; modifications of grazing procedures; and extended use of cover cropping, the use of noneconomic crops to improve the quality of pasturelands (Gerber, Henderson, and Makkar 2013).

Mining Wastes

Mining is a very large industry in the United States. In 2019, more than four dozen important minerals were mined in this country, some in very large amounts, others in smaller quantities (table 2.3).

All mining operations result in the production of waste materials. The amount of wastes generated depends on the mineral being mined and the method used for mining. In general, four types of mining are in use. *Underground mining* involves boring into the earth to some depth to reach the mineral. Underground coal mines are among the most familiar of such structures. They usually consist of vertical shafts that extend hundreds or thousands of feet into the ground, with horizontal tunnels extending outward from the vertical shaft. *Surface mining* includes open-pit mining, strip mining, quarrying, and other forms of mineral recovery that involve the removal of surface material in order to reach the mineral to be mined. This form of mining can have devastating effects on the region in which resource extraction occurs (see, for example, http://beniciaindependent.com/yes-they-are-bomb-trains-even-more-importantly-they-are-global-destruction-trains/). *Placer mines* are operations in which minerals are recovered from streams, and *in situ mining* involves the extraction of a mineral by physical or chemical means without trying to extract it from the ground (Copeland, n.d.).

Waste production occurs at every stage of resource extraction by mining. The first stage of such processes involves the removal of as much earth, rock, and other materials as may be necessary to get at the resource. In coal mining, for example, very large equipment may be used to scrape topsoil and underlying materials off the surface to make it possible for strip mining to begin to occur. The materials that are removed in this process are called *overburden*, *spoil*, or simply *waste*. The amount of overburden for any particular mining operation depends on a host of factors, one of which may be the amount of resource available. In strip mining for coal, for example, it

Table 2.3 Mine Production of Selected Minerals in the United States, 2014–2018 (in thousand metric tons*)

Mineral	2014	2015	2016	2017	2018
Aluminum	1,710	1,587	818	741	890
Beryllium	0.270	0.205	0.155	0.150	0.170
Coal**	1,000,049	896,941	728,364	774,609	755,523
Cobalt	0.120	0.760	0. 690	0.640	0.500
Copper	1,360	1,380	1,430	1,260	1,200
Gold	0.210	0.214	0.228	0.237	0.210
Gypsum	18,300	18,800	19,800	20,700	21,000
Iron ore	56,100	46,100	41,800	47,900	49,000
Lead	378	370	346	310	260
Lime	19,500	18,300	17,700	17,800	19,000
Molybdenum	68.2	47.4	36.2	40.7	42.0
Nickel	4.3	27.2	24.1	22.1	19.0
Phosphate rock	25,300	27,400	27,100	27,900	27,000
Palladium	0.012	0.012	0.013	0.013	0.014
Platinum	0.037	0.037	0.039	0.040	0.041
Potash	850	740	510	480	500
Salt	45,300	45,100	41,700	40,000	42,000
Silver	1.180	1.09	1.14	1.03	0.90
Zeolites	62.8	75.1	75.2	82.4	95.0

*One metric ton = 1.1 short tons.

**Thousand short tons.

Source: "Mineral Commodity Summaries 2019." 2019. U.S. Geological Survey. https://doi.org/10.3133/70202434; http://prd wret.s3 us west 2.amazonaws.com/assets/palladium/production/atoms/files/mcs2019_all.pdf. Accessed on July 19, 2019. For coal: "U.S. Coal Summary Statistics, 2013–2019." U.S. Energy Information Administration. https://www.eia.gov/coal/production/quarterly/pdf/tes1p01p1.pdf. Accessed on July 19, 2019.

may be necessary to remove only a relatively thin layer of material to get at the coal itself. In such cases, the ratio of overburden to resource can be fairly small. For less abundant materials, such as gold or silver, it may be necessary remove relatively large amounts of overburden to get to and begin extracting the resource. It is not uncommon for the volume of overburden to be as much as ten times the volume of the resource recovered. (An illustrative estimate of overburden in one mining region can be found at "NorthMet Project Rock and Overburden Management Plan" 2015.)

In most cases, the amount of overburden to be dealt with can be very large, hundreds or thousands of cubic yards at least. Transporting those wastes to some distant location is generally not viable economically. Companies may simply move the wastes to some accessible location nearby where they will have (hopefully) minimal environmental impact. (An image of an overburden deposition site is available at "Valley Fills" 2013.) In some cases, overburden may be used to remediate a mining site when operations end there. In fact, some states and regions *require* that some sort of remediation occur, with overburden a good candidate material to use in the area. Such actions typically do not occur, however, as an act of a company's goodwill. That option is generally too expensive for most companies to take on voluntarily. (A photograph of an overburden area that has been remediated can be found at Hardy 2015.)

A second stage of waste production occurs when the desired resource is separated from the material in which it is embedded. For example, a small sliver of gold metal may be embedded within a mixture of clay, sand, and other rocky material. The economically nonuseful material removed from a mine is called the *tailings*. The material has a host of other names also, including *culm dumps*, *mine dumps*, *leach residue*, *slickens*, *slimes*, *tails*, or *terra-cone*. For most of human history, miners and mining companies have simply taken the easiest way out for mine tailings: They have dumped them into nearby rivers, lakes, or other waterways. The problem with this system, as one might guess, is that tailings often contain materials that are toxic to plants, humans, and other animals. One of the best-known examples of toxic tailings is illustrated by iron-ore mining. When pure iron ore is extracted from mine deposits, it may exist in the form of or be contaminated by compounds of sulfur, such as pyrite (FeS_2) and iron(II) sulfide (FeS). When exposed to air and water, these compounds are converted into a class of compounds known as sulfates, the most harmful of which is sulfuric acid (H_2SO_4). Reports of "acid mine drainage," then, are occasioned by rivers and other waterways that have been contaminated by levels of sulfuric acid that make them uninhabitable by aquatic organisms and toxic to humans and other land animals (Coil et al. 2014).

Mining waste has generally been excluded from federal legislation dealing with hazardous and nonhazardous wastes. Over time, the U.S. Congress has taken a variety of actions to specifically protect the mining industry from having to follow provisions of waste management established by the Resource Conservation and Recovery Act of 1976 and the Solid Waste Disposal Act of 1980. After a very long battle between environmentalists and industry lobbyists, the Environmental Protection Agency finally published in 1987 a long and detailed document, "Mining Waste Exclusion," that defines in great detail mining wastes that will not be covered by federal legislation and the conditions under which other wastes will be so monitored. As with other federal regulations, however, it is not clear how the EPA or other agencies might change these conditions at some time in the future. In the meanwhile, states and local governments have tended to be much more aggressive in the regulation of mining wastes ("Special Wastes" 2018; for the Mining Waste Exclusion rule itself and its history, see "Mining Waste" 2016; "Mining Waste Exclusion" 1989/2014; as an example of state regulation, see "State Mining and Geology Board Statutes and Regulations" 2019).

An important exception to the exclusion of mine wastes from regulatory action is related to the dumping of mine tailings in waterways. One of the most contentious of those actions has been the so-called Stream Protection Rule. That rule was first proposed in 1983 by the U.S. Office of Surface Mining, Reclamation, and Enforcement. In its original form, the rule forbade the dumping of mining wastes within one hundred feet of a river or stream. The rule was never enforced very strongly, and efforts continued over the next three decades to make it more effective. Finally, in December 2016, the administration of President Barack Obama was able to announce a much stronger version of the rule. That action did not last very long, however, as newly elected president Donald Trump had campaigned on elimination of the rule. The U.S. Congress passed legislation abandoning the Obama rule in February 2017, and mining companies are, as of late 2019, still allowed to dump toxic tailings into the nation's waterways (Plumer 2017).

A common theme in the modern application of waste management systems to mining involves a hierarchical plan that begins with *waste avoidance*. As with other waste management schemes, the most desirable step is usually to try producing a smaller amount of waste, thus reducing the problems involved in disposing of those wastes. One example of this approach makes use of the so-called mill cut-off grade. This term refers to the point a company sets at which mining for a mineral is no longer economically profitable. At that point, some mineral may remain in wastes, but mining ends for economic reasons. By lowering the cut-off grade, mining can continue and reduce the amount of wastes produced in the process ("Environmentally Sensitive 'Green' Mining" 2016; Hancock 2018; for example, Prasetya and Simatupang 2012).

Another approach for reducing the amount of mining wastes in the environment is underground storage. Deposition in deep wells has long been proposed and used as a method of dealing with hazardous wastes of many types. Recently, greater attention has been given to the possibility of using the technology for dealing with hazardous mine tailings ("Waste Storage 'In Perpetuity'" 2012). Mine waste management can also involve the reuse of wastes for new purposes. For example, numerous studies have been conducted on the use of mine tailings for road construction; manufacture of bricks, tiles, pellets, and other ceramic materials; production of paints; fabrication of plastic wood; and other industrial uses ("Use of Tailings," n.d.; User Guidelines for Waste and Byproduct Materials in Pavement Construction 2016).

Neutralization of hazardous substances in mining wastes has long been conducted by chemical means. The appropriate chemicals selected for these purposes are used to react with, precipitate, or otherwise act on toxic substances in wastes to reduce or eliminate their hazardous properties. More recently, another approach to the treatment of hazardous mine wastes has been with the use of biological agents. Several microorganisms have been discovered that attack and degrade hazardous substances at less cost and with greater efficiency than traditional chemical agents. (See, for example, Johnson 2014; for a good overview of the field of mine waste management, see Lottermoser 2010.)

Environmental Justice

One topic that has received relatively little attention in this book so far is the siting of solid wastes—that is, what happens when a city or town decides that it needs a new solid wastes disposal site? After all, no one wants to have a smelly dump or landfill, a busy waste transfer station, or an incinerator spewing toxic gases into the air across from his or her home. That means that the most common answer to the problem of siting is NIMBY, or "not in my back yard." So even people who are supportive of and enthusiastic about a new waste disposal facility are almost certain to demand that it be built "anywhere but in my neighborhood."

A major criterion for building a new solid waste disposal facility then is that it be placed on land not currently in use or occupied only by individuals who are not in a strong position to fight back against the new facility. This means that a new solid waste disposal facility is most likely to end up in areas occupied by people of color or low-income communities.

A classic example of this situation occurred in the city of Houston in 1978 when the city granted a permit to Southwestern Waste Management, Inc., to build a new sanitary landfill, called the Whispering Pines Sanitary Landfill, in the Northwood Manor subdivision of the city. Residents of the area, which was then 82 percent black, objected to the siting decision and filed suit against the company. They hired attorney Linda McKeever Bullard to prosecute the case, which came to be called *Bean v. Southwestern Land Management*. Bullard, in turn, hired her husband, Robert, to collect data that would support her case. Robert Bullard was, at the time, assistant professor of sociology at Texas Southern University. He and his students studied waste disposal sites in Houston at the time and found that all five of the city-owned landfills, three out of four privately owned landfills, and six out of eight city incinerators were located in black neighborhoods. Overall, at a time when blacks made up 25 percent of the city's population, 82 percent of all waste and garbage dumped in Houston ended up in black neighborhoods ("An Early Legal Challenge to LULU Sitings: Northwood Manor, Texas" 1998; "Invisible Houston: Full Interview with Dr. Robert Bullard, Father of Environmental Justice Movement" 2017).

(Ms. Bullard lost *Bean v. Southwestern*. The court agreed that siting a solid waste disposal site close to a residential area that included a high school without air conditioning "did not make sense." But it concluded that the company had no "intent to discriminate on the basis of race" and was not, therefore, guilty of environmental racism [Northern 1997, 538].)

The Houston example of disproportionate exposure to solid waste siting was hardly unique, or even unusual. Over succeeding years, similar examples were uncovered by researchers, described in the literature, and often the subject of additional legal action (Northern 1997, 539, et. seq.; Newton 2009,

6–18). The tendency of individuals, organizations, and communities to fight back against such policies and practices has come to be known as the *environmental justice* movement. The term *environmental justice* has been defined as "the fair treatment and meaningful involvement of all people regardless of race, color, national origin, or income, with respect to the development, implementation, and enforcement of environmental laws, regulations, and policies" ("Environmental Justice" 2019).

An important breakthrough in the environmental justice movement occurred in 1987 with the publication of a report by the Commission for Racial Justice of the United Church of Christ. That report summarized the results of two studies conducted by a special committee on racial inequities in the siting of waste disposal facilities in the United States. It reported five major conclusions from the research (quoting from the report):

- Race proved to be the most significant among variables tested in association with the location of commercial hazardous waste facilities. This represented, a consistent national pattern.
- Communities with the greatest number of commercial hazardous waste facilities had the highest composition of racial and ethnic residents.
- In communities with two or more facilities or one of the nation's five largest landfills, the average minority percentage of the population* was more than three times that of communities without facilities (38 percent vs. 12 percent).
- In communities with one commercial hazardous waste facility, the average minority percentage of the population was twice the average minority percentage of the population in communities without such facilities (24 percent vs. 12 percent).
- Although socio-economic status appeared to play an important role in the location of commercial hazardous waste facilities, race still proved to be more significant. This remained true after the study controlled for urbanization and regional

differences. Incomes and home values were substantially lower when communities with commercial facilities were compared to communities in the surrounding counties without facilities.

- Three out of the five largest commercial hazardous waste landfills In the United States were located in predominantly Black* or Hispanic communities. These three landfills accounted for 40 percent of the total estimated commercial landfill capacity in the nation.

*In this report, "minority percentage of the population" was used as a measure of "race." ("Toxic Wastes and Race in the United States" 1987, xiii–xiv)

At least in part because of the United Church of Christ report, the environmental justice movement had reached a turning point in the 1990s. From a relatively small, often ineffective, effort located in discrete, local communities, the movement had become a national campaign to address the disproportionate exposure of people of color and lower incomes to offensive and hazardous waste. For example, a group of academics and activists met at the University of Michigan in 1990 in the first national conference on environmental justice. A group formed at the meeting, the so-called Michigan Coalition, issued a report on the status of environmental inequities, "Race and the Incidence of Environmental Health." They also sent a letter to William Reilly, administrator of the EPA, in which they "demand[ed] action on environmental risks in minority and low-income communities and on tribal lands." In response to the letter, Reilly appointed the Environmental Equity Workgroup (EEG). The EEG was made up of members from all EPA offices and regions across the country. It was charged with the task of "assess[ing] the evidence that racial minority and low-income communities bear a higher environmental risk burden than those in the general population, and consider[ing] what EPA might do about any identified disparities" ("Environmental Equity. Reducing Risk for All Communities" 1992, 2). Two years later, EEG was renamed the Office of Environmental Justice, an agency that remains in effect today.

So, how much has changed with regard to environmental inequities since the late 1980s? In 2007, the Justice and Witness Ministries of the United Church of Christ commissioned a group of four experts in the field of environmental justice to assess changes that had taken place in the preceding twenty years. The four had been active in the environmental justice movement from its earliest days and reviewed and carried out research to see how it had developed over two decades. They concluded that "[t]wenty years after the release of [the report] Toxic Wastes and Race, significant racial and socioeconomic disparities persist in the distribution of the nation's commercial hazardous waste facilities." They explained that they had used "newer methods that better match where people and hazardous waste facilities are located," but those methods produced results that "are very much the same as they were in 1987" (Bullard et al. 2007). A number of reports appear to have confirmed that conclusion, not only in 2007, but in the two decades since the most recent report. (See, for example, Baptista and Amarnath 2016; Martuzzi, Mitis, and Forastiere 2010; Norton 2019; "Recycling and Environmental Justice," n.d.)

Another retrospective report on the evolution of the environmental justice movement was published in 2014. That report reviewed developments in the movement between 1964 and 2014. Its primary conclusion appeared to be that environmental justice was no longer a fringe topic, primarily of interest to individual communities with their own unique issues. Instead, the movement had become a well-known and highly appreciated national phenomenon characterized by environmental justice laws in all fifty states; more extensive research leading to better policy decisions; a system for developing new leaders for the movement; stable academic programs and university-based centers; and national awards, honors, and other forms of recognition for participants in the field. In other words, environmental justice had become part of the political establishment. In spite of these changes, however, it was unclear precisely how much actual improvement had been made for populations historically affected by environmental injustices (Bullard et al. 2014).

There is some evidence that questions about the environmental justice move still exist. In 2019, for example, the EPA announced that it was changing a rule that allowed individuals and communities to appeal agency rulings on the curtailment of pollution by certain types of facilities (Davenport 2019). The right to appeal has for many years been a powerful tool for those concerned with environmental equity and for those who want to reduce pollution to which they might be exposed. Efforts must still be made, obviously, to convince some policymakers of the need for guaranteeing that people of color and poor communities not bear an unequal burden from the inappropriate disposal of all types of waste products.

References

Abbing, Michiel Roscam. 2019. *Plastic Soup: An Atlas of Ocean Pollution*. Washington, DC: Island Press.

"Agricultural Waste Management Systems." 2011. U.S. Department of Agriculture. Agricultural Waste Management Field Handbook. https://directives.sc.egov.usda.gov/OpenNonWebContent.aspx?content=31493.wba. Accessed on July 15, 2019.

"Algae Basics." n.d. All about Algae.com. http://allaboutalgae.com/benefits/. Accessed on July 16, 2019.

"All You Need to Know about Algae Biofuel." 2017. The Earth Project. https://theearthproject.com/algae-biofuel/. Accessed on July 16, 2019.

"Ammonia." 2019. Environmental Protection Agency. https://www.epa.gov/caddis-vol2/ammonia. Accessed on July 16, 2019.

Awasthi, Abhishek Kumar, et al. 2019. "Circular Economy and Electronic Waste." *Nature Electronics*. 2(3): 86–89.

Baldé, C. P., et al. 2017. "The Global E-waste Monitor—2017." United Nations University, International Telecommunication Union, and International Solid Waste Association. https://www.itu.int/en/ITU-D/Climate-Change/Documents/GEM%202017/Global-E-waste%20Monitor%202017%20.pdf. Accessed on July 13, 2019.

Baptista, Ana Isabel, and Kumar Kartik Amarnath. 2016. "Garbage, Power, and Environmental Justice. The Clean Power Plan Rule." *William and Mary Environmental and Law Policy Review*. 41(2): 403–433. https://scholarship.law.wm.edu/cgi/viewcontent.cgi?referer=https://www.google.com/&httpsredir=1&article=1675&context=wmelpr. Accessed on July 21, 2019.

Benbrook, Charles M. 2016. "Trends in Glyphosate Herbicide Use in the United States and Globally." *Environmental Sciences Europe*. 28(1): 1–15. https://enveurope.springeropen.com/articles/10.1186/s12302-016-0070-0. Accessed on July 17, 2019.

Bennett, Michael. 2019. "New Study Highlights States, Cities Most At-Risk from Toxic Roundup Weed Killer." Weed Killer Crisis. https://www.weedkillercrisis.com/topics/most-and-least-pesticides-america/#top20. Accessed on July 17, 2019.

"Biopesticide Active Ingredients." 2018. Environmental Protection Agency. https://www.epa.gov/ingredients-used-pesticide-products/biopesticide-active-ingredients. Accessed on July 17, 2019.

Botequilha-Leitão, André. 2012. "Industrial Ecosystem at Kalundborg, Denmark." https://www.researchgate.net/figure/The-several-components-of-the-industrial-eco-system-at-Kalundborg-Denmark-and-its_fig3_234016076. Accessed on July 14, 2019.

Bruckner, Monica. 2018. "The Gulf of Mexico Dead Zone." Microbial Life. https://serc.carleton.edu/microbelife/topics/deadzone/index.html. Accessed on July 16, 2019.

Bullard, Robert D., et al. 2014. "Environmental Justice: Milestones and Accomplishments, 1964–2014." Barbara

Jordan-Mickey Leland School of Public Affairs. Texas Southern University. http://www.racialequitytools.org/resourcefiles/Enviromental_justice.pdf. Accessed on July 21, 2019.

Bullard, Robert D., et al. 2007. "Toxic Wastes and Race at Twenty 1987–2007." Justice and Witness Ministries. United Church of Christ. https://www.nrdc.org/sites/default/files/toxic-wastes-and-race-at-twenty-1987-2007.pdf. Accessed on July 21, 2019.

"China—World's Dumping Ground for Electronic Waste (CNN)." 2013. https://www.youtube.com/watch?v=O-_ubuFhqQA. Accessed on July 13, 2019.

Chua, Jasmin Malik. 2019. "Basically Everything You're Given on an Airplane Is Wrapped in Plastic. That's a Huge Environmental Problem." Vox. https://www.vox.com/the-goods/2019/7/9/20680969/airplanes-plastic-zero-waste-flights. Accessed on July 11, 2019.

Coil, David, et al. 2014. "Acid Mine Drainage." Ground Truth Trekking. http://www.groundtruthtrekking.org/Issues/MetalsMining/AcidMineDrainage.html. Accessed on July 19, 2019.

Copeland, Rudolph. n.d. "Non-Renewable Mineral Resources." Slide Player. https://slideplayer.com/slide/7062829/. Accessed on July 19, 2019.

"Corn Planted Acreage Up 3 Percent from 2018. Corn Stocks Down 2 Percent from June Last Year." 2019. U.S. Department of Agriculture. Natural Agricultural Statistics Service. https://www.nass.usda.gov/Newsroom/2019/06-28-2019.php. Accessed on July 15, 2019.

Davenport, Carol. 2019. "E.P.A. Plans to Curtail the Ability of Communities to Oppose Pollution Permits." *The New York Times*. https://www.nytimes.com/2019/07/12/climate/epa-community-pollution-appeal.html. Accessed on July 21, 2019.

"Defining Hazardous Waste: Listed, Characteristic and Mixed Radiological Wastes." 2019. Environmental Protection Agency. https://www.epa.gov/hw/defining-hazardous-waste-listed-characteristic-and-mixed-radiological-wastes. Accessed on July 13, 2019.

"Disposal of High-Level Nuclear Waste." 2019. U.S. Government Accountability Office. https://www.gao.gov/key_issues/disposal_of_highlevel_nuclear_waste/issue_summary#t=0. Accessed on July 11, 2019.

Dunkley, C. S., and K. D. Dunkley. 2013. "Greenhouse Gas Emissions from Livestock and Poultry." *Agriculture, Food and Analytical Bacteriology*. 3: 17–29. https://pdfs.semanticscholar.org/eb03/2436d1063bca66bc9a94949f912773ddfebd.pdf. Accessed on July 17, 2019.

"An Early Legal Challenge to LULU Sitings: Northwood Manor, Texas." 1998. Stanford Law School Environmental and Natural Resources Law and Policy Program. http://media.law.stanford.edu/organizations/programs-and-centers/enrlp/doc/slspublic/northwood.pdf. Accessed on July 21, 2019.

"Eco-Industrial Park Initiative for Sustainable Industrial Zones in Viet Nam." 2018. United Nations Industrial Development Organization. http://eipvn.org/wp-content/uploads/2016/10/EIP-Brochure.pdf. Accessed on July 14, 2019.

"Eco-Industrial Parks (EIP)." 2006. Indigo Development. http://www.indigodev.com/Ecoparks.html. Accessed on July 14, 2019.

"Eco-Industrial Parks & Industrial Ecology." 2018. Making Lewes. https://makinglewes.org/category/eco-industrial-parks/. Accessed on July 14, 2019.

"Environmental Equity. Reducing Risk for All Communities." 1992. Environmental Protection Agency. https://nepis.epa.gov/Exe/ZyPDF.cgi/40000JLA.PDF?Dockey=40000JLA.PDF. Accessed on July 21, 2019.

"Environmental Justice." 2019. Environmental Protection Agency. https://www.epa.gov/environmentaljustice. Accessed on July 20, 2019.

"Environmentally Sensitive 'Green' Mining." 2016. The Future of Strategic Natural Resources. https://web.mit.edu/12.000/www/m2016/finalwebsite/solutions/greenmining.html. Accessed on July 20, 2019.

"EPA Approves Expanded Use of Indiana-Made Insecticide Toxic to Bees." 2019. Indiana Environmental Reporter. http://indianaenvironmentalreporter.org/posts/epa-approves-expanded-use-of-indiana-made-insecticide-toxic-to-bees. Accessed on July 18, 2019.

"Ethanol Timeline." 2011. Historic Vehicle Association. https://www.historicvehicle.org/ethanol-timeline/. Accessed on July 15, 2019.

"E-waste Facts We All Need to Know." 2019. American Recycling. https://americanrecyclingne.com/e-waste-facts-we-all-need-to-know/. Accessed on July 12, 2019.

"Fighting for Trash-Free Seas." 2019. Ocean Conservancy. https://oceanconservancy.org/trash-free-seas/plastics-in-the-ocean/. Accessed on July 11, 2019.

"Garbage in China." 2019. Facts and Details. http://factsanddetails.com/china/cat10/sub66/item1111.html#chapter-17. Accessed on July 12, 2019.

Gelfand, Ilya, and G. Philip Robertson. 2015. "Mitigation of Greenhouse Gas Emissions in Agricultural Ecosystems." In S. K. Hamilton, J. E. Doll, and G. P. Robertson, eds. *The Ecology of Agricultural Landscapes: Long-Term Research on the Path to Sustainability*, 310–339. New York: Oxford University Press. https://pdfs.semanticscholar.org/1a23/ed5b833caf919f56d300f25b395aeba8932b.pdf. Accessed on July 17, 2019.

Gerber, Pierre J., Benjamin Henderson, and Harinder P. S. Makkar, eds. 2013. "Mitigation of Greenhouse Gas

Emissions in Livestock Production: A Review of Technical Options for Non-CO_2 Emissions." Rome: Food and Agriculture Organization of the United Nations. https://www.uncclearn.org/sites/default/files/inventory/fao180.pdf. Accessed on July 17, 2019.

"Global Mitigation of Non-CO_2 Greenhouse Gases: 2010–2030." 2013. Environmental Protection Agency. https://www.epa.gov/sites/production/files/2016-06/documents/mac_report_2013.pdf. Accessed on July 17, 2019.

Gould, Kevin. 2007. "Corn Stover Harvesting." Beef Brief. Michigan State University. https://www.canr.msu.edu/uploads/236/58572/CornStoverHarvesting.pdf. Accessed on July 15, 2019.

Gramling, Carolyn. 2009. "As Green as It Gets: Algae Biofuels." *Earth*. https://www.earthmagazine.org/article/green-it-gets-algae-biofuels. Accessed on July 16, 2019.

Grant, Richard. 2009. "Drowning in Plastic: The Great Pacific Garbage Patch Is Twice the Size of France." *The Telegraph*. https://www.telegraph.co.uk/news/earth/environment/5208645/Drowning-in-plastic-The-Great-Pacific-Garbage-Patch-is-twice-the-size-of-France.html. Accessed on July 11, 2019.

"Great Pacific Garbage Patch." 2019. National Geographic. https://www.nationalgeographic.org/encyclopedia/great-pacific-garbage-patch/. Accessed on July 11, 2019.

"Greenhouse Gas Emissions from Composting of Agricultural Wastes." 2001. Alberta Agriculture, Food and Rural Development. https://www1.agric.gov.ab.ca/$department/deptdocs.nsf/all/cl3014/$file/GHGBulletinNo6Composting.pdf?OpenElement. Accessed on July 17, 2019.

Grossi, Giampero, et al. 2019. "Livestock and Climate Change: Impact of Livestock on Climate and Mitigation Strategies." *Animal Frontiers*. 9(1): 69–76. https://academic.oup.com/af/article/9/1/69/5173494. Accessed on July 17, 2019.

Guenthner, Joseph F., et al. 1999. "Assessment of Pesticide Use in the U.S. Potato Industry." *American Journal of Potato Research*. 76(1): 25–29. https://www.researchgate.net/publication/257217768_Assessment_of_pesticide_use_in_the_US_potato_industry. Accessed on July 17, 2019.

"A Guide for Industrial Waste Management." 2006. Environmental Protection Agency. https://www.epa.gov/sites/production/files/2016-03/documents/industrial-waste-guide.pdf. Accessed on July 14, 2019.

Hancock, Tracy. 2018. "Mining Waste Set to Grow, but 'Reduce, Reuse, Recycle' Solutions Abound." *Mining Weekly*. https://www.miningweekly.com/article/mining-waste-set-to-grow-but-reduce-reuse-recycle-solutions-abound-2018-09-07-1. Accessed on July 20, 2019.

Hardy, Nathaniel. 2015. "That's Not a Real Hill—Hunter Valley Coal Mining." SESL Australia. https://sesl.com.au/blog/that-s-not-a-real-hill-hunter-valley-coal-mining/. Accessed on July 19, 2019.

Harris, Ebrahim, Rozanna Latiff, and Robert Birsel. 2019. "Malaysia to Send 3,000 Tonnes of Plastic Waste Back to Countries of Origin." SWI. https://www.swissinfo.ch/eng/malaysia-to-send-3-000-tonnes-of-plastic-waste-back-to-countries-of-origin/44993318. Accessed on July 11, 2019.

Heaton, Emily A., et al. 2019. "Miscanthus (Miscanthus x giganteus) for Biofuel Production." Farm Energy. https://farm-energy.extension.org/miscanthus-miscanthus-x-giganteus-for-biofuel-production/. Accessed on July 15, 2019.

Hou, Yong, Gerard L Velthof, and Oene Oenema. 2015. "Mitigation of Ammonia, Nitrous Oxide and Methane Emissions from Manure Management Chains: A Meta-analysis and Integrated Assessment." *Global Change Biology*. 21(3): 1293–1312.

"Information for Exporters of Resource Conservation and Recovery Act (RCRA) Hazardous Waste." 2019. Environmental Protection Agency. https://www.epa.gov/

hwgenerators/information-exporters-resource-conservation-and-recovery-act-rcra-hazardous-waste#Summary. Accessed on July 10, 2019.

"International Agreements on Transboundary Shipments of Hazardous Waste." 2019. Environmental Protection Agency. https://www.epa.gov/hwgenerators/international-agreements-transboundary-shipments-hazardous-waste. Accessed on July 10, 2019.

"Invisible Houston: Full Interview with Dr. Robert Bullard, Father of Environmental Justice Movement." 2017. Democracy Now. https://www.democracynow.org/2017/9/7/invisible_houston_full_interview_with_dr. Accessed on July 21, 2019.

Jaffe, Daniel. 1995. "The International Effort to Control the Transboundary Movement of Hazardous Waste: The Basel and Bamako Conventions," *ILSA Journal of International & Comparative Law*. 2(1): 123–137. https://core.ac.uk/download/pdf/51088942.pdf. Accessed on July 10, 2019.

Johnson, D. Barrie. 2014. "Recent Developments in Microbiological Approaches for Securing Mine Wastes and for Recovering Metals from Mine Waters." *Minerals*. 4: 279–292. https://doi.org/10.3390/min4020279. https://www.mdpi.com/2075-163X/4/2/279. Accessed on July 20, 2019.

Johnson, Lisa K., et al. 2018. "Field Measurement in Vegetable Crops Indicates Need for Reevaluation of On-farm Food Loss Estimates in North America." *Agricultural Systems*. 167: 136–142.

Johnson, Jane M.-F., et al. 2007. "Agricultural Opportunities to Mitigate Greenhouse Gas Emissions." *Environmental Pollution*. 150: 107–124. https://pubag.nal.usda.gov/download/15176/PDF. Accessed on July 17, 2019.

Kleppel, Gary. 2014. *The Emergent Agriculture Farming, Sustainability and the Return of the Local Economy*. Gabriola Island, BC: New Society Publishers.

"Legal Limits on Single-Use Plastics and Microplastics: A Global Review of National Laws and Regulations." 2018. United Nations Environment Programme. https://wedocs.unep.org/bitstream/handle/20.500.11822/27113/plastics_limits.pdf?sequence=1&isAllowed=y. Accessed on July 11, 2019.

Liboiron, Max. 2016. "Municipal versus Industrial Waste: Questioning the 3–97 Ratio." Discard Studies. https://discardstudies.com/2016/03/02/municipal-versus-industrial-waste-a-3-97-ratio-or-something-else-entirely/. Accessed on July 14, 2019.

" 'A Line in the Sand'—Ellen MacArthur Foundation Launches New Plastics Economy Global Commitment to Eliminate Plastic Waste at Source." 2018. Ellen MacArthur Foundation. https://www.ellenmacarthurfoundation.org/news/a-line-in-the-sand-ellen-macarthur-foundation-launch-global-commitment-to-eliminate-plastic-pollution-at-the-source. Accessed on July 11, 2019.

"Livestock's Long Shadow [LLS]." 2006. Livestock, Environment and Development Initiative. http://www.fao.org/docrep/010/a0701e/a0701e.pdf. Accessed on July 17, 2019.

Lottermoser, Bernd G. 2010. *Mine Wastes: Characterization, Treatment and Environmental Impacts*, 3rd ed. Berlin: Springer. http://kamceramics.com/wp-content/uploads/2017/02/Bernd_G._Lottermoser_Mine_Wastes_CharacterizatiBookZZ.org_.pdf. Accessed on July 20, 2019.

Lowe, Ernest. 2015. "A First Tool of Industrial Ecology: Eco-industrial Parks." Dartmouth University. http://www.dartmouth.edu/~cushman/courses/engs171/EIPs.pdf. Accessed on July 14, 2019.

"Low-level Radioactive Waste (LLW)." 2019. United State Nuclear Regulatory Commission. https://www.nrc.gov/

reading-rm/basic-ref/glossary/low-level-radioactive-waste-llw.html. Accessed on July 11, 2019.

Malkan, Stacy. 2019. "Glyphosate Fact Sheet: Cancer and Other Health Concerns." U.S. Right to Know. https://usrtk.org/pesticides/glyphosate-health-concerns/. Accessed on July 17, 2019.

"Managing Industrial Solid Wastes from Manufacturing, Mining, Oil and Gas Production, and Utility Coal Combustion." 1992. Office of Technology Assessment. https://ota.fas.org/reports/9225.pdf. Accessed on July 14, 2019.

"Marine Plastic Pollution." 2019. Ocean United. https://www.oceanunite.org/issues/marine-plastic-pollution/. Accessed on July 11, 2019.

Martuzzi, Marco, Francesco Mitis, and Francesco Forastiere. 2010. "Inequalities, Inequities, Environmental Justice in Waste Management and Health." *European Journal of Public Health*. 20(1): 21–26. https://doi.org/10.1093/eurpub/ckp216. https://academic.oup.com/eurpub/article/20/1/21/611240. Accessed on July 21, 2019.

"Meat Statistics Recent." 2019. U.S. Department of Agriculture. Economic Research Service. https://www.ers.usda.gov/data-products/livestock-meat-domestic-data/livestock-meat-domestic-data/#All%20meat%20statistics. Accessed on July 15, 2019.

"The Memo." 1991. The Whirled Bank Group. http://www.whirledbank.org/ourwords/summers.html. Accessed on July 10, 2019.

Meylan, Frédéric, et al. 2017. United Nations Industrial Development Organization. https://www.unido.org/sites/default/files/files/2018-05/UNIDO%20Eco-Industrial%20Park%20Handbook_English.pdf. Accessed on July 14, 2019.

Miner, Reid. 2010. *Impact of the Global Forest Industry on Atmospheric Greenhouse Gases*. Rome: Food and Agriculture

Organization of the United Nations. https://www.researchgate.net/publication/237420638_The_effects_of_the_global_forest_products_industry_on_atmospheric_greenhouse_gases_FAO_Forestry_Paper_159. Accessed on July 17, 2019.

"Mining Waste." 2016. Environmental Protection Agency. https://archive.epa.gov/epawaste/nonhaz/industrial/special/web/html/index-5.html#history. Accessed on July 20, 2019.

"Mining Waste Exclusion." 1989/2014. *Federal Register*. 54(169): 36592–36642. https://www.epa.gov/sites/production/files/2016-03/documents/54fr36592.pdf. Accessed on July 20, 2019.

Moore, Charles. 2014. *Plastic Ocean: How a Sea Captain's Chance Discovery Launched a Determined Quest to Save the Oceans*. New York: Avery.

Muralikrishna, Iyyanki V., and Valli Manickam. 2017. "Hazardous Waste Management." In Iyyanki V. Muralikrishna and Valli Manickam. *Environmental Management: Science and Engineering for Industry*, chapter 17. Oxford, UK; Cambridge, MA: Elsevier; Butterworth-Heinemann. https://www.sciencedirect.com/topics/earth-and-planetary-sciences/hazardous-waste-management. Accessed on July 14, 2019.

Nance, Trevor. 2018. "Here's a List of Every City in the US to Ban Plastic Bags, Will Your City Be Next?" http://www.akleg.gov/basis/get_documents.asp?session=31&docid=24240. Accessed on July 11, 2019.

Nathanson, Jerry A. 2019. "Hazardous-Waste Management." *Encylopædia Britannica*. https://www.britannica.com/technology/hazardous-waste-management. Accessed on July 11, 2019.

"National Overview: Facts and Figures on Materials, Wastes and Recycling." 2018. Environmental Protection Agency. https://www.epa.gov/facts-and-figures-about-materials-waste-and-recycling/national-overview-facts-and-figures-materials#Generation. Accessed on June 28, 2019.

"A New Circular Vision for Electronics: Time for a Global Reboot." 2019. World Economic Forum. http://www3.weforum.org/docs/WEF_A_New_Circular_Vision_for_Electronics.pdf. Accessed on July 13, 2019.

"Newport Aquarium Joins National Push to Take #FirstStep to Cut Plastic Pollution." 2018. Newport Aquarium. https://aquariumworks.org/tag/no-straw-november/. Accessed on July 11, 2019.

Newton, David E. 2009. *Environmental Justice: A Reference Handbook*, 2nd ed. Santa Barbara, CA: ABC-CLIO.

Ni, Hong-Gang, and Eddy Y. Zeng. 2009. "Law Enforcement and Global Collaboration Are the Keys to Containing E-Waste Tsunami in China." *Environmental Science and Technology*. 43: 3991–3994. https://pubs.acs.org/doi/pdf/10.1021/es802725m?rand=jmdcljc6. Accessed on July 12, 2019.

Niranjan, Ajit. 2019. "Amid Plastic Deluge, Southeast Asia Refuses Western Waste." DW. https://www.dw.com/en/amid-plastic-deluge-southeast-asia-refuses-western-waste/a-49467769-0. Accessed on July 10, 2019.

Northern, Kathy Seward. 1997. "Battery and Beyond: A Tort Law Response to Environmental Racism." *William & Mary Environmental Law and Policy Review*. 21(3): 485–598. https://scholarship.law.wm.edu/cgi/viewcontent.cgi?article=1289&context=wmelpr. Accessed on July 21, 2019.

"NorthMet Project Rock and Overburden Management Plan." 2015. Barr Engineering Co. https://www.leg.state.mn.us/docs/2015/other/150681/PFEISref_2/PolyMet%202015h.pdf. Accessed on July 19, 2019.

Norton, Jennifer M. 2019. "Environmental Injustice, Public Health and Solid Waste Facilities in North Carolina." https://cdr.lib.unc.edu/concern/dissertations/dz010q80r. Accessed on July 21, 2019.

Panoutsou, Calliope, and Asha Singh. 2017. "D4.1 Training Materials for Agronomists and Students." Panacea. http://www.panacea-h2020.eu/wp-content/uploads/2019/05/D4.1-Training-manual-for-agronomists-and-students_INI-format-review_as.pdf. Accessed on July 15, 2019.

Parker, Laura. 2018. "China's Ban on Trash Imports Shifts Waste Crisis to Southeast Asia." National Geographic. https://www.nationalgeographic.com/environment/2018/11/china-ban-plastic-trash-imports-shifts-waste-crisis-southeast-asia-malaysia/. Accessed on July 10, 2019.

Paustian, Keith, et al. 2004. "Agricultural Mitigation of Greenhouse Gases: Science and Policy Options." Washington, DC: Conference on Carbon Sequestration. https://pdfs.semanticscholar.org/4ad1/c374924b9269a267ce516fa4a32826a465ca.pdf. Accessed on July 17, 2019.

Perkins, Devin N., et al. 2014. "E-waste: A Global Hazard." *Annals of Global Health*. 80(4): 286–295. https://reader.elsevier.com/reader/sd/pii/S2214999614003208?token=AB7CDB3B8926694ECCF6B7C8EE519B3B463F1E810BBF4D71777921F78577715D708641CD2B22F3A8BF674158C12ADD0C. Accessed on July 12, 2019.

"Philadelphia Ash Dumping Chronology." 2008. http://faculty.webster.edu/corbetre/haiti-archive/msg03569.html. Accessed on July 10, 2019.

Plumer, Brad. 2017. "Why Trump Just Killed a Rule Restricting Coal Companies from Dumping Waste in Streams." Vox. https://www.vox.com/2017/2/2/14488448/stream-protection-rule. Accessed on July 19, 2019.

Pomranz, Mike. 2018a. "Carlsberg Solved the Six-Pack Packaging Problem with Glue." Food & Wine. https://www.foodandwine.com/news/carlsberg-snap-pack-glue. Accessed on July 11, 2019.

Pomranz, Mike. 2018b. "Corona to Test Plastic-Free Six-Pack Rings on Cans." Food & Wine. https://www.foodandwine.com/beer/corona-plastic-free-biodegradable-six-pack-rings-test. Accessed on July 11, 2019.

Popp, József, Károly Pető, and János Nagy. 2013. "Pesticide Productivity and Food Security. A Review." *Agronomy for Sustainable Development*. 33(1): 243–255. https://link.springer.com/content/pdf/10.1007%2Fs13593-012-0105-x.pdf. Accessed on July 17, 2019.

Prasetya, Lukman Budi, and Togar M. Simatupang. 2012. "Cut-Off Grade Optimization at Grasberg Surface Mine in Considering Environmental Impact." *Proceedings of the 3rd International Conference on Technology and Operations Management: Sustaining Competitiveness through Green Technology Management*. 89–106. https://www.academia.edu/1858305/Cut-off_Grade_Optimization_at_Grasberg_Surface_Mine_in_Considering_Environmental_Impact. Accessed on July 20, 2019.

"Public Health Statement for NITRATE and NITRITE." 2015. Agency for Toxic Substances and Disease Registry. https://www.atsdr.cdc.gov/phs/phs.asp?id=1448&tid=258. Accessed on July 16, 2019.

"Recycling and Environmental Justice." n.d. Green America. https://www.greenamerica.org/rethinking-recycling/recycling-and-environmental-justice. Accessed on July 21, 2019.

"Resource Conservation and Recovery Act (RCRA) Laws and Regulations." 2019. Environmental Protection Agency. https://www.epa.gov/rcra. Accessed on July 13, 2019.

"ReVenture Park—Charlotte's First Eco-Industrial Park." 2014. ReVenture Park. http://www.reventurepark.com/. Accessed on July 14, 2019.

Rindels, Michelle, and Humberto Sanchez. 2019. "Federal Government Discloses It Already Shipped Plutonium to

Nevada, Without State's Knowledge or Consent." *The Nevada Independent*. https://thenevadaindependent.com/article/federal-government-discloses-it-already-shipped-plutonium-to-nevada-without-states-knowledge-or-consent. Accessed on July 11, 2019.

Ritchie, Hannah, and Max Roser. 2018. "Plastic Pollution." Our World in Data. https://ourworldindata.org/plastic-pollution#plastic-waste-per-person. Accessed on July 10, 2019.

Ryder, Guy, and Houlin Zhao. 2019. "The World's E-waste Is a Huge Problem. It's Also a Golden Opportunity." World Economic Forum. https://www.weforum.org/agenda/2019/01/how-a-circular-approach-can-turn-e-waste-into-a-golden-opportunity/. Accessed on July 13, 2019.

Schmidt, Charles W. 1999. "Trading Trash: Why the U.S. Won't Sign on to the Basel Convention." *Environmental Health Perspectives*. 107(8): A410–A412. https://www.ncbi.nlm.nih.gov/pmc/articles/PMC1566487/pdf/envhper00513-0030-color.pdf. Accessed on July 10, 2019.

Schwartz, Jerry. 2000. "The Full Story of the Khian Sea and the Gonaives Ash Mountain." http://faculty.webster.edu/corbetre/haiti-archive-new/msg05049.html. Accessed on July 10, 2019.

"70% of Annual Global E-waste Dumped in China." 2012. China.org.cn. http://www.china.org.cn/environment/2012-05/24/content_25461996.htm. Accessed on July 12, 2019.

"Single-Use Plastics: A Roadmap for Sustainability." 2018. United Nations Environmental Programme. https://wedocs.unep.org/bitstream/handle/20.500.11822/25496/singleUsePlastic_sustainability.pdf? Accessed on July 11, 2019.

Smith, Matt. 2017. "Trump Eyes Rebooting Yucca Mountain, as Nuclear Waste Piles Up." Seeker. https://www.seeker.com/earth/energy/trump-eyes-rebooting-yucca-mountain-as-nuclear-waste-piles-up. Accessed on July 11, 2019.

Smith, Pete, et al. 2008. "Greenhouse Gas Mitigation in Agriculture." *Philosophical Transactions of the Royal Society of London. Series B, Biological Sciences.* 363(1492): 789–813. https://www.ncbi.nlm.nih.gov/pmc/articles/PMC2610110/. Accessed on July 17, 2019.

"Solving the E-waste Challenge Requires Global Action." 2019. Phys.org. https://phys.org/news/2019-03-e-waste-requires-global-action.html. Accessed on July 13, 2019.

"Special Wastes." 2018. Environmental Protection Agency. https://www.epa.gov/hw/special-wastes#mining. Accessed on July 20, 2019.

"State Mining and Geology Board Statutes and Regulations." 2019. California Department of Conservation. https://www.conservation.ca.gov/smgb/Regulations. Accessed on July 20, 2019.

Staub, Colin. 2018. "Export Market Roundup: Vietnam Bans E-plastic Imports." E-Scrap News. https://resource-recycling.com/e-scrap/2018/12/06/export-market-roundup-vietnam-bans-e-plastic-imports/. Accessed on July 10, 2019.

Steinfeld, Henning, et al. 2006. "Livestock's Long Shadow: Environmental Issues and Options." Rome: Food and Agriculture Organization of the United Nations. http://www.fao.org/3/a-a0701e.pdf. Accessed on July 16, 2019.

Swift, Peter. 2015. "The Proposed Yucca Mountain Repository: A Case Study." Sandia National Laboratories. https://www.osti.gov/servlets/purl/1332844. Accessed on July 11, 2019.

Takahashi, Jun'ichi. and Bruce A Young, eds. 2001. "Greenhouse Gases and Animal Agriculture." *Proceedings of the 1st International Conference on Greenhouse Gases and Animal Agriculture*, Obihiro, Japan, 7–11 November, 2001. Amsterdam; Boston: Elsevier.

Thompson, James. 2018. "Time Is Running Out: The U.S. Landfill Capacity Crisis." Solid Waste Environmental

Excellence Protocol. https://nrra.net/sweep/time-is-running-out-the-u-s-landfill-capacity-crisis/. Accessed on July 10, 2019.

"Timeline: 1954–2016." 2019. Yucca Mountain.org. https://www.yuccamountain.org/time.htm. Accessed on July 11, 2019.

"Tips to Use Less Plastic." 2018. Green Education Foundation. http://www.greeneducationfoundation.org/nationalgreenweeksub/waste-reduction-tips/tips-to-use-less-plastic.html. Accessed on July 11, 2019.

"Toxic Wastes and Race in the United States." 1987. Commission for Racial Justice. United Church of Christ. https://www.nrc.gov/docs/ML1310/ML13109A339.pdf. Accessed on July 21, 2019.

Tsamis, Achilleas, and Mike Coyne. 2015. "Recovery of Rare Earths from Electronic wastes: An Opportunity for High-Tech SMEs." Directorate General for Internal Policies. Policy Department A: Economic and Scientific Policy. http://www.europarl.europa.eu/RegData/etudes/STUD/2015/518777/IPOL_STU(2015)518777_EN.pdf. Accessed on July 13, 2019.

"2016 Billion-Ton Report: Advancing Domestic Resources for a Thriving Bioeconomy." 2016. Department of Energy. https://www.energy.gov/sites/prod/files/2016/12/f34/2016_billion_ton_report_12.2.16_0.pdf. Accessed on July 15, 2019.

"U.S. Department of Energy's Motion to Withdraw." 2010. United States of America Nuclear Regulatory Commission. https://www.yuccamountain.org/pdf/doe_motion_to_withdraw-2010_nrc_app.pdf. Accessed on July 11, 2019.

"Underground Injection Control (UIC)." 2016. Environmental Protection Agency. https://www.epa.gov/uic/class-iv-shallow-hazardous-and-radioactive-injection-wells. Accessed on July 12, 2019.

"Use of Tailings." n.d. Samarco. https://www.samarco.com/en/aproveitamento-de-rejeitos/. Accessed on July 20, 2019.

"User Guidelines for Waste and Byproduct Materials in Pavement Construction." 2016. Federal Highway Administration Research and Technology. https://www.fhwa.dot.gov/publications/research/infrastructure/pavements/97148/037.cfm. Accessed on July 20, 2019.

Vadas, P. A., et al. 2009. "Estimating Phosphorus Loss in Runoff from Manure and Fertilizer for a Phosphorus Loss Quantification Tool." *Journal of Environmental Quality*. 38(4): 1645–1653. https://pdfs.semanticscholar.org/de11/1148691c5c2e995251da95ea92b3f35d8a92.pdf. Accessed on July 17, 2019.

"Valley Fills." 2013. Kids Love Mountains. https://kidslovemountains.wordpress.com/learn-more-2/valley-fills/. Accessed on July 19, 2019.

Van Bruggen, A. H. C., et al. 2018. "Environmental and Health Effects of the Herbicide Glyphosate." *Science of the Total Environment*. 616–617: 255–268.

Verma, Rinku, et al. 2016. "Toxic Pollutants from Plastic Waste—A Review." *Procedia Environmental Sciences*. 35: 701–708. https://reader.elsevier.com/reader/sd/pii/S187802961630158X. Accessed on July 11, 2019.

"Volumes & Amounts of Manure Produced by Livestock." 2018. http://agrienvarchive.ca/bioenergy/facts.html#Volumes_&_Amounts. Accessed on July 15, 2019.

Wang, Wanli, et al. 2019. "Current Influence of China's Ban on Plastic Waste Imports." *Waste Disposal & Sustainable Energy*. 1(1): 67–78. https://link.springer.com/article/10.1007/s42768-019-00005-z. Accessed on July 10, 2019.

Ward, Mary H. 2009. "Too Much of a Good Thing? Nitrate from Nitrogen Fertilizers and Cancer." *Reviews on Environmental Health*. 24(4): 357–363. https://www.ncbi.nlm.nih.gov/pmc/articles/PMC3068045/. Accessed on July 16, 2019.

"Waste Storage 'In Perpetuity.'" 2012. Ground Truth Trekking. http://www.groundtruthtrekking.org/Issues/OtherIssues/perpetual-waste-storage-perpetuity.html. Accessed on July 20, 2019.

"Wastes." 2018. Environmental Protection Agency. https://www.epa.gov/report-environment/wastes. Accessed on July 12, 2019.

"What Is a Circular Economy?" 2017. Ellen MacArthur Foundation. https://www.ellenmacarthurfoundation.org/circular-economy/concept. Accessed on July 13, 2019.

"Wheat Data." 2019. U.S. Department of Agriculture. Economic Research Service. https://www.ers.usda.gov/data-products/wheat-data/. Accessed on July 15, 2019.

Zhou, Naaman. 2019. "Malaysia to Send up to 100 Tonnes of Plastic Waste Back to Australia." *The Guardian*. https://www.theguardian.com/world/2019/may/29/malaysia-to-send-up-to-100-tonnes-of-plastic-waste-back-to-australia. Accessed on July 10, 2019.

Zimring, Carl A., and William L. Rathje, eds. 2012. *Encyclopedia of Consumption and Waste: The Social Science of Garbage*. Thousand Oaks, CA: SAGE Publications.

Introduction

Waste management is not only a topic of scientific, technical, economic, social, political, and other interest but also one about which individuals have a variety of opinions. The essays in this chapter discuss some of the technical issues involved in waste management as well as their personal viewpoints on some of these issues.

Computational Approaches to Precise Toxicology Research

Hussain Ather

The lack of knowledge about toxicity raises questions of how to preserve the basic human rights to breathe clean air, drink clean water, and eat clean food. The lack of toxicology knowledge costs the United States about $5–$8 billion a year, said James Glazier, director of the Indiana University Biocomplexity Institute. There are about 100,000 chemicals that generate the household products people use every day, but scientists only know the basic toxicological effects of about 4–6 percent of them, according to Joseph Shaw, associate professor in the School of Public and Environmental Affairs at Indiana University Bloomington (Ather 2016).

A gardener uses a shovel to turn over garden compost made from food scraps. (Reinout Van Wagtendonk/Dreamstime.com)

Assessing waste management's effect on health means understanding how its toxins specifically target various parts of the human body. For most chemicals in different household items, their direct harmful effects on human health are relatively unknown. Toxicology researchers can embrace basic biological science and physiology to study the effects of hazardous chemicals on human health to improve the time, precision, and cost efficiency of toxicology research in general. Computational techniques such as machine learning can aid researchers in drawing new conclusions by harnessing the power of big data.

The field of toxicology seeks to determine the harmful effects that chemicals and other substances have on people and the environment. Researchers can form predictions on which chemicals cause harm and how they affect the body.

The research would revolve around basic questions of how humans may be exposed to toxicological threats and figuring out their effects on people. Though researchers may have general ideas of how exposure to harmful compounds can cause mild symptoms like skin rashes or even diseases like cancer, research needs to embrace more efficient testing methods to assess and predict all relevant health effects. Improving tests from being slow and costly to being more effective and individualized to specific chemicals would give scientists a greater insight into what happens at the physiological level when people are exposed to harmful compounds.

Shaw studies genetics and cellular mechanisms behind how organisms respond to the environment to provide the basis for how chemicals may influence people. Drawing data from experimental science and public health alongside computational approaches to analyze large amounts of data, Shaw and Glazier together can assess thousands of molecules at once while reenvisioning toxicology research as scientists know it. The National Institutes of Health consider the research a first step in creating technologies to predict therapeutic agent toxicity as well as toxins that may pollute the environment. The two use computer simulations to assess toxins for effects on the liver in test organisms such as mice.

Other computational approaches to toxicology have used machine learning methods to form predictions on the hazardous effects of chemicals and other substances. Yunyi Wu and Guanyu Wang, researchers in the biology department of the Southern University of Science and Technology in Shenzhen, China, analyzed toxicity prediction using acute-toxicity data from the Hazardous Substances Data Bank and Toxicity ForeCaster. With recent advances in deep learning methods of forming predictions from data that is unstructured and sparse, yet great in number, the researchers converted molecules into undirected graphs that correspond to their chemical features such as the location of specific functional groups. From there, they can deduce trends in classifying and predicting based on properties like molecular weight, degradation rate, solubility coefficient in different solvents, molar index, and permeability. The researchers borrowed methods of classifying toxicity between acute and chronic toxicity, the latter of which affects reproduction, mutagenicity, and carcinogenicity (Wu and Wang 2018).

Liu et al. (2018) created gene interaction networks such that toxicity can correlate with the connectedness in the network to capture the complexity of late-onset, chronic toxicity. Using a combination of chemical structure and change in genetic expression, the team created a nearest-neighbor model to predict the quantitative relationship between the structure and expression. For acute toxicity, Wu and Wang created a prediction method that involved processing a chemical structure using a convolutional kernel that operated at the level of the individual atom to create pre-fingerprints that combined to create a deep-minded fingerprint. The two used a regression and classification model to test the predictions. They showed the deep learning approach was more effective in handling larger amounts of data than previously established random forest approaches.

These machine learning approaches let scientists avoid making costly, harmful, and risky experimental trials when they are not necessary. They let researchers harness the power of large amounts of data using an approach that combines transcriptome and genome data with molecular chemical information about compounds and molecules themselves. Wu and Wang even argue machine learning methods may become entirely governed by computational methods as data and methods increase in efficiency and effectiveness.

References

Ather, Hussain. 2016. "SPEA Professor Re-envisions Toxicology Research." *Indiana Daily Student.* https://www.idsnews.com/article/2016/09/spea-professor-re-envisionstoxicology-research. Accessed on July 10, 2019.

Liu Ruifeng, et al. 2018. "Assessing Deep and Shallow Learning Methods for Quantitative Prediction of Acute Chemical Toxicity." *Toxicological Sciences.* 164(2): 512–526. https://academic.oup.com/toxsci/article/164/2/512/4990897. Accessed on July 10, 2019.

Wu, Yunyi, and Guanyu Wang. 2018. "Machine Learning Based Toxicity Prediction: From Chemical Structural Description to Transcriptome Analysis." *International Journal of Molecular Sciences.* 19(8): 2358. https://doi.org/10.3390/ijms19082358. https://www.ncbi.nlm.nih.gov/pmc/articles/PMC6121588/. Accessed on July 10, 2019.

Hussain Ather is a researcher-writer in neuroscience and philosophy. He is completing his master's of science in science communication at the University of California Santa Cruz and hopes to pursue a PhD in computational neuroscience.

The Recycling Puzzle
Sandy Becker

The title of this volume is "Waste Management." The best "management" plan would be to eliminate "waste" completely—use everything up! That's easier said than done, so let's focus on how to recycle or compost as much as we can.

My Connecticut town has "single stream recycling." That means we put paper, cans, glass, and plastic all into one bin. Then the stuff is picked up by the waste hauler and taken to a mechanical sorting facility, which makes further choices about what can really be "recycled"—that is, processed and turned back into something useful and therefore profitable—and what is a "contaminant" (we shouldn't have put it in the recycling bin in the first place).

I was inspired to write this essay after going to a recycling workshop here in my hometown, with a population of a little over 47,000. In short, it was exasperating. The rules for what we can and can't recycle here in Connecticut are really, really complicated! The machinery that sorts our single stream of recycled material is extremely fussy about what it can handle. Therefore, the rules about what is "acceptable" in the eyes of the recycling facility are really, really complicated. The presenter handed out 8½" × 11" sheets listing what's OK and not OK to put in the recycle bin. For example, paper is acceptable, but shredded paper isn't—because somehow the sorting machines can't handle shredded paper. Ice cream containers, although they are paper, aren't OK because they have some kind of waxy coating. Yet milk cartons are fine. Glass and plastic bottles are acceptable, but their caps aren't (unless screwed on securely)—again, loose caps gum up the sorting machinery. Glass jars are acceptable, but glass drinking glasses aren't (that one wasn't explained by the presenter). Single-use plastic cups are OK, but single-use paper cups aren't. And so on.

I wasn't the only one in the audience muttering to my neighbor that the recycling haulers should get better sorting machines. This audience of about a hundred consisted of people pretty committed to doing a responsible job of recycling their waste. That's why we went to the workshop. Yet many of us came away so indignant at the complexity of the rules that we probably wouldn't follow them very well. So what could we expect of the 46,900 neighbors who didn't even come to the workshop?

Compared to Europeans, we Americans are pretty bad at dealing with our waste products. For example, according to the World Economic Forum website, Germans recycle or compost 56 percent of their waste, while Americans only recycle or compost 34 percent (Gray 2017). The rest goes in the landfill. Why are we so bad at this? Well, it turns out European countries, and many Asian countries, have rather different recycling procedures, with economic incentives and cultural expectations in place to encourage citizens to toe the line.

To get an idea of how it works in some other countries, I talked to an old friend who had lived in Germany for many years. He said that in Munich, households have a bin for paper and one for table scraps, which are picked up every couple of weeks. Scattered around town are bigger bins to which householders take their metal, plastic, and glass for recycling. He mentioned that people in Munich expect themselves and their neighbors to do this sorting and binning responsibly. The data suggests that they do.

I also checked with a couple who live in the United Kingdom. They sort recycling into four categories: paper, cardboard and plastics, metals and glass, food and garden waste. Containers are picked up from households every two weeks and go to a nearby processing facility. Again, citizens, not machines, are doing the initial sorting.

Notice that in these examples food and garden waste are picked up and turned into compost. Not many people in this country compost their food waste and grass clippings, partly because they have no use for the resulting compost, but if this stuff were picked up by the local recycling collectors, people might put it out separately, instead of throwing it away with the trash. I recently got an email notice that our town dump is now accepting compostable food waste—but you have to cart it to the dump yourself! How many citizens are going to do that?

I talked to our local recycling coordinator, and she confirmed that while single stream recycling is easier for the haulers to collect, and for people to put out, it's harder to separate into usable streams. The result is that some things can't be recycled at all—like shredded paper and food waste. And the finicky sorting machine will sometimes send truly recyclable items to the landfill because it can't separate them from the trash that someone incorrectly threw in the recycling bin.

I began this essay grumbling about the unreasonable complexity of my local recycling rules. How can I be expected to do a good job when the rules are so messy and arbitrary? But as I began writing, and searching out more information, I've come to suspect that the problem may be that we have a culture that resists "rules." Europeans sort their recyclable items into several bins, so they don't need much of a sorting machine. They collect food scraps for composting. They cart their bottles and cans and plastics to the appropriate neighborhood bins. Are Americans kind of lazy? It really sounds like Germans and Brits just work harder at recycling than we do.

My daughter lives in Vermont, the most eco-friendly of the fifty states (Breyer 2018). Her family collects recyclable paper, plastic, and glass and takes it all to a collection facility. They have a compost pile in the backyard. As of next summer, food waste will be banned from the landfills, so everyone will have to either compost it or take it to a composting center ("Waste Not Guide" n.d.). So there's hope for us yet.

References

Breyer, Melissa. 2018. "The 10 Eco-friendliest States in the US." Treehugger. https://www.treehugger.com/environmental-policy/top-10-eco-friendly-states-us.html. Accessed on August 29, 2019.

Gray, Alex. 2017. "Germany Recycles More Than Any Other Country." World Economic Forum. https://www.weforum.org/agenda/2017/12/germany-recycles-more-than-any-other-country/. Accessed on August 29, 2019.

"Waste Not Guide." n.d. Vermont. https://dec.vermont.gov/sites/dec/files/wmp/SolidWaste/Documents/VT-Waste-Not-Guide.pdf. Accessed on August 29, 2019.

Sandy Becker was a cell biologist for several decades and is now retired. She moonlights as a science writer, covering biology in the broadest sense, environmental issues, and economics (which in her view are all connected).

The Search for Safe and Effective Wastewater Management Solutions

Adrienne Fung

Cesspools have a history that is several centuries old, improving on the practice of chamber pots and sewage gutters that used to line the streets. Recently, however, cesspools have been in the spotlight, seen as harmful problems or misunderstood scapegoats.

Controversial Cesspools

Those who see cesspools as problems point to the negative impacts that cesspools have on groundwater, caused by the leaching of cesspool contents into the surrounding area ("Cesspools in Hawaii" 2019). If contaminated groundwater discharges into a nearby body of water, this could potentially affect recreational users or fishers, or animal and plant life. Or, if the groundwater is pumped up from a drinking water well, cesspool contamination could affect human health.

These images, as argued by those who see cesspools as a misunderstood party, rely on misguided assumptions that cesspools do little or nothing to treat raw wastewater. Bacteria may play a role in biological treatment and decomposition, and the surrounding soil could provide filtration for the liquid waste (Hawaii Pacific Engineers 1999, 7–2 to 7–4). Negligent maintenance has also led to problems that have been blamed solely on cesspools. Moreover, cesspools provide an economical and convenient source of nitrogen and phosphorus as fertilizer.

Need for Change

Today, billions of dollars are being spent on improving onsite sewage disposal systems, primarily cesspools. In 2009, the State of Florida, Department of Health, contracted a $5 million study that included evaluating and recommending reliable and cost-effective alternatives to cesspools. Up at West Falmouth in Massachusetts, up to $10,000 was given to each homeowner as a subsidy in a project to upgrade cesspools.

The movement is also occurring through legislature, for example, with the U.S. Environmental Protection Agency (EPA)'s 2000 ban on new construction of large-capacity cesspools, followed by a 2005 ban on existing large-capacity cesspools. On a more local level, states have also enacted cesspool laws in different degrees. In Hawaii, replacements of all cesspools in the state are required to be completed by 2050. Starting in 2019, residents of Suffolk County in New York will have to meet minimum treatment requirements when replacing cesspools.

The vast amounts of resources directed toward cesspools stem from human health concerns, particularly nitrate/nitrite contamination. Leachate from cesspools is a known contributor of nitrate/nitrite to the environment. Through nitrification, nitrite is converted to nitrate in the presence of oxygen, and since nitrate is more stable, this compound is more commonly found.

Nitrate in drinking water, however, can lead to adverse health effects in young children by interfering with oxygen transport in the bloodstream. This is known as methemoglobinemia or blue baby syndrome and can be caused by water with nitrate at as little as twelve milligrams per liter (mg/L) (Knobeloch 2000, 675–678). For protection of human health, the U.S. EPA established a maximum contaminant level of 10 mg/L for nitrate (as nitrogen) in drinking water. To remove harmful nitrogen, denitrification under anoxic conditions reduces nitrate to gaseous nitrogen (Hazen and Sawyer 2009, 2–4).

This occurs to a limited extent in systems relying on cesspools; therefore, many cesspool alternatives are based on promoting denitrification.

Moving Forward

As one can imagine, as well intentioned as the cesspool bans or mandatory cesspool upgrades are, questions arise regarding costs, whether the new systems will really solve the issues, or whether cesspools are even that harmful. These and other questions are valid concerns that are on the minds of homeowners, who will ultimately carry most of the weight for financing and maintaining their converted systems. Lawmakers also face pressure from differing parties who either defend cesspools or push for upgrades. For a smoother transition, there are three focus areas that deserve attention: law, research, and outreach.

Law: Enacting Change

Much of change is driven top-down through laws. As such, it would be in society's best interest for government entities to guide and assist compliance with cesspool-related laws. A comprehensive plan would be instrumental, including milestones and specific steps to reach them. Additionally, incentives for upgrading cesspools could be implemented through subsidies, grants, loans, or tax credits. Prioritization of areas for cesspool conversions would also be helpful by allocating resources to places most in need of upgrades.

Research: Building Knowledge

Second, research is crucial for improved understanding of both cesspools and other onsite sewage treatment and disposal systems as well as for innovating wastewater management solutions that better protect human health and the environment. Major considerations of cesspool upgrades include available space, geology, topography, and soil type, as there is no one-size-fits-all solution. Therefore, it is necessary to have a range of different options. Moreover, since proper functioning is key to any system, there has recently been a lot of growth in passive wastewater treatment and disposal systems that reduce the requirements for regular maintenance (Hazen and Sawyer 2015).

Outreach: Taking Action

Even with laws in effect and research in growth, the action for anything to be done lies with each individual. Outreach is of utmost importance to inform the community of human health and environmental motivations for cesspool conversions and to provide resources for guiding and supporting individuals through the process. Listening to the perspectives of homeowners, community members, and other stakeholders is going to shape the paths of lawmakers and researchers.

Overall, a lot of work and questions remain in solving the cesspools issue, but the goal is safer, cleaner drinking water that will benefit everyone.

References

"Cesspools in Hawaii." 2019. State of Hawaii Department of Health. https://health.hawaii.gov/wastewater/cesspools/. Accessed on August 31, 2019.

Hawaii Pacific Engineers. 1999. "Final Submittal Waimanalo Wastewater Facilities Plan." Honolulu: City and County of Honolulu, Department of Design and Construction.

Hazen and Sawyer, Environmental Engineers and Scientists. 2015. "Evaluation of Full Scale Prototype Passive Nitrogen Reduction Systems (PNRS) and Recommendations for Future Implementation." Florida Department of Health. http://www.floridahealth.gov/environmental-health/onsite-sewage/research/_documents/rrac/hazensawyervoli reportrmall.pdf. Accessed on August 31, 2019.

Hazen and Sawyer, Environmental Engineers and Scientists. 2009. "Literature Review of Nitrogen Reduction Technologies for Onsite Sewage Treatment Systems." Florida Department of Health. http://www.floridahealth.gov/environmental-health/onsite-sewage/research/_documents/nitrogen/task-a-lit-review.pdf. Accessed on August 31, 2019.

Knobeloch, L., et al. 2000. "Blue Babies and Nitrate-Contaminated Well Water." *Environmental Health Perspectives*. 108(7): 675–678.

Adrienne Fung is an environmental engineer specializing in drinking water quality and remediation projects. She has a bachelor of science in engineering degree in chemical and biological engineering from Princeton University and a master of science in civil engineering from the University of Hawaii. Ms. Fung serves in leadership roles in the American Water Works Association and the Hawaii Association of Environmental Professionals, helping to connect people to work together and solve water issues.

Ecological Feces and Urine Alchemy in Shogun Japan

Joel Grossman

After centuries of internal warfare and severe environmental degradation, Shogun Japan, ruled by former samurai warriors from 1600 to 1868, isolated itself from the world to heal and recover, and created a sustainable, sanitary society notable for monetizing and ecologically recycling human feces and urine as night-soil fertilizers that boosted crop yields to feed an expanding population. Keeping sewage out of waterways also meant less pollution, healthier seafood, and a much lower incidence of plague and disease than London or Paris, where rivers doubled as sewage dumps and drinking water sources.

Edo, modern Tokyo's precursor, was a cluster of small fishing villages when the Tokugawa Dynasty selected it for the new

Shogun capital in 1590. By 1700, Edo (Tokyo) had mushroomed to a city of 1 million, roughly double the size of London and Paris. Aqueducts and wooden pipes delivered healthful, unpolluted fresh spring and river water to Edo. London boasted metal pipes and modern water closets but flushed feces into the River Thames where drinking water was harvested. In contrast to Japanese cities such as Tokyo and Osaka, London suffered centuries of devastating cholera and typhoid fever outbreaks from drinking water polluted with sewage.

In 1853, after U.S. Commodore Matthew Perry's "gunboat diplomacy" opened Japan to expanded trade and foreign visitors, Europe and North America began getting firsthand reports of relatively clean Japanese cities with decorative toilets watched over by a toilet god promoting bumper crops. Recycling toilet wastes was a nationwide industry with guilds and associations and employed thousands in what by all accounts was a smelly profession. Although Louis Pasteur's germ theory of disease causation did not gain acceptance until the late 1800s, centuries earlier the Japanese practiced sanitation because they internalized the Shinto religion's emphasis on cleanliness and purification rituals. Also, in the Far East (China, Korea, Japan), night soil traditionally had high value as farm fertilizer, so dumping it into waterways was like throwing away silver or gold.

Integrating human waste into food production systems predated ecological science and the environmental movement by centuries. By some accounts, human excrement was Japan's third most valuable commodity behind rice and cotton. Very early in the 1600s, the Shogun and an army of former samurai serving as bureaucrats strictly enforced regulations abolishing all habitations along rivers, ending sewage runoff into waterways and ensuring healthy fisheries.

In marketplaces, farmers bartered commodities like daikon radishes, eggplants, and pickles for feces and urine to fertilize fields. In apartment buildings, landlords traditionally owned the feces and tenants owned the urine, which fetched a lower price because it had fewer nutrients and was more difficult to transport. The more tenants per apartment, the lower the rent, as landlords collected more feces. On average, excrement from twenty people bought enough grain to feed one person for a year.

Farmers were notorious for building guest toilets and along with entrepreneurs schemed and fought legal battles to place urinals and toilets on busy urban street corners and along remote rural roadways. In contrast, the 17th-century Palace of Versailles was built without toilets. The French king used a chamber pot, often in front of honored guests, while courtiers relieved themselves under stairwells, along corridors, in courtyards, and against secluded buildings. And try finding a public toilet in a 21st-century American city.

Japanese poop buyers became shrewd connoisseurs, recognizing centuries before the discovery of nitrogen that the diets of wealthy people produced the most nutrient-rich excrement. As night soil demand exceeded supply, prices inflated, leading to bidding wars and even theft. Annual contracts often set prices in silver, but gold was required to buy feces from royalty. In 1908, a decade before Fritz Haber's chemistry Nobel Prize for the Haber-Bosch process synthesizing ammonia from atmospheric nitrogen, Japan fertilized its fields with 24 million tons of human manure, roughly 1.75 tons per acre. Modern Japan replaced human manure with 400,000 metric tons of synthetic nitrogen, which at $500–$700 per ton costs $200–$300 million, not to mention the high cost of sewage treatment plants.

In 1908, the United States and Europe were "pouring into the sea, lakes or rivers and into the underground waters" six to twelve million pounds of nitrogen per million people per year, along with vast quantities of phosphorous and potassium, "and this waste we esteem one of the great achievements of our civilization," wrote Francis King (King 1911, 194). "In the Far East, for more than thirty centuries, these enormous wastes have been religiously saved."

"When waste has a positive value, people aren't going to throw it away, and so excreta was not dumped into the streets, as it was in European cities, nor was it allowed to seep into the ground" via cesspools, wrote Susan Hanley (Hanley 2001, 42). "In some cities maids emptied chamber pots out windows, and streets in London had open sewers running down the middle of them as late as the early-18th century. Even in the 1880s, Cambridge, England, was described as 'an undrained, river-polluted cesspool city'." Streets in American cities were no better, and possibly worse. In 1857, streets in New York were described as "one mass of wreaking, disgusting filth," phraseology still useful for some downtown areas in the United States.

"The night soil collection 'industry' was made possible and profitable by gigantic networks of collection, fermentation and composting, and distribution," wrote Anthony Walsh (Walsh 2009, 64). "Enormous barges were employed to ship the night soil, while tankers transported urine all the way to the cotton fields." By boiling water for tea, cooking vegetables, and removing shoes inside homes, Shinto Japan had germ protection. Unfortunately, numerous factors, including a post–World War II U.S. occupation army averse to "industry" smells and with different eating habits succumbed to germs, spurring sewage "modernization" and abandonment of ecological recycling.

References

Hanley, Susan B. 2001. "Advanced Public Sanitation in 17th–19th Century Japan." *Journal of Japanese Trade and Industry*. https://www.jef.or.jp/journal/pdf/unknownjapan_0101.pdf. Accessed on August 3, 2019.

King, Francis H. 1911. *Farmers of Forty Centuries*. Madison, WI: Mrs. F. H. King.

Walsh, Anthony. 2009. "Economies of Excrement: Public Health and Urban Planning in Meiji Japan." *Historical Perspectives: Santa Clara University Undergraduate Journal of History* (Series II). 14(1): 55–76. http://scholarcommons.scu.edu/historical-perspectives/vol14/iss1/9. Accessed on August 3, 2019.

Joel Grossman worked with Los Angeles mayor Tom Bradley's office on a central composting project for horse manure and bedding and writes the Biocontrol Control Beat blog.

Colonel Waring's Epoch Cure

Jim Nordlinger

As it turned out, George E. Waring Jr., a nationally celebrated and remarkably influential self-made sanitary engineer, had the science of infection, in an era of rampant contagious epidemics, significantly and, for him, tragically wrong.

But driven by great confidence, especially in his misunderstanding of contagion's cause, supported by an outstanding national reputation-based achievement and salesmanship and a remarkable set of skills as a technical and popular writer, fiscal manager and fundraiser, innovator and civic executive, he carried out a remedy for an unimaginable, even vile, waste management ill and got the cure, at least for an epidemic of Gilded Age inequality and the filthy streets of 1895 New York City, miraculously right.

George Waring, born July 4, 1833, in Pound Ridge, New York, was a farmer, who trained in the chemistry and science of agriculture in nearby Poughkeepsie. He polished his reputation by lecturing around New England. While managing Horace Greeley's farm, he gained the admiration of Frederick Law Olmstead. Part of the original group that created New York's Central Park, George Waring designed its water and drainage systems. In 1861, he joined the Union cavalry in the Civil War and was promoted from major to colonel (Cassedy 1962, 165).

After the war, without formal medical training, laboratory experience, or practice, Colonel Waring became a sanitary engineer and an entrepreneur of public health. Even for the

American medical community, he became a leading anticontagionist, arguing against Pasteur and Koch's germ theory, loudly asserting that filth is the environment that engenders the spontaneous development of the agents of infectious disease. He published numerous pamphlets and influential books, such as *Draining for Profit* and *Draining for Health* and later *The Sanitary Condition of City and Country Dwelling Houses.* He was a propagandist for the removal of filth and proper drainage, based on technologies in which he invested (Cassedy 1962, 164).

Beyond his self-promotion, Colonel Waring proved himself to be a remarkably effective man of action and a showman. Summoned by the U.S. government, first as part of a panel, and then alone, he cleaned up the sewage and drainage systems of 1878 Memphis, Tennessee, after a series of devastating yellow fever epidemics that made him nationally renowned. Remaining staunchly anticontagionist, he became a wealthy man (Cassedy, 1962, 165).

But the 1890s' New York City was different. The challenge of cleaning up the astoundingly accumulated, deep, and sprawling filth in miles of streets, neglected for years, was not simply one for the science of contagion. The city's population, having grown steadily between 1850 and 1890, more than doubled from 1890 to 1900, from about 1.5 million to more than 3.4 million, from a torrent of immigration.

Unfathomably filth laden, much of the city's scourge was political and social, generated by an ethos of purposefully corrupt negligence, carried out by Tammany Hall: years of jobs exchanged for votes and copiously siphoned funds. But it was also the Gilded Age, a chasm of weltering, angry disparity between the wealthy few and the vast number of tenement-dwelling desperate poor, with one manifest difference: the wealthy neighborhoods had privately funded sanitation service. The poor had corrupt Tammany Hall or, it seemed, no one. But no person was spared the danger of the buried streets, clogged with refuse, 60,000 abandoned carts and wagons, the carcasses of putrefying horses and cats, animal and human excrement, discarded mattresses thrown from windows, endless paper, old food, and a thick layer of ash, which became paste when it rained.

In 1894, William Lafayette Strong, banker and old New York Republican, who had run for elected office once and lost, was elected the city's mayor, on a surge of public outrage in response to the revelations of lurid Tammany Hall scandals (Nagle 2013, 104).

Colonel Waring, Strong's second choice to lead the Department of Street Cleaning (DSC, renamed the Department of Sanitation in 1929), after Theodore Roosevelt, future president and eminent Rough Rider, had the important backing of the Ladies' Health Protective Association, who wanted civic cleanliness. He accepted Mayor Strong's offer as long as "I get my own way." When the mayor suggested that the1881 charter prevented that request, Colonel Waring reportedly answered, "I don't mean the law; I mean you." Strong could remove him, Colonel Waring asserted, but he could not interfere (Nagle 2013, 104).

On March 29, 1895, little more than three months after being sworn in, Colonel Waring announced, very publically at a meeting of the Good Government Club, he had an effective plan: "It is to put a man instead of a voter at the end of the broom handle. When I took charge of the department I found the men discouraged in the belief that they would be turned out and replaced by good Republicans but the men have learned now that it doesn't matter what their politics may be; if they attend to their work they will be kept" ("A Man at the End of a Broom" 1895).

While Colonel Waring enacted many innovations—separation of refuse, the forerunner of recycling; special bags for ash collection; a new rolling cart; fiscal innovations—he also engendered controversy. Against the union, he tried to hire immigrant women as cheap labor. Still, the greatest focus was on the worker.

He reorganized the city's districts into smaller sections, cleanable in a day. Each morning, sweeper-carter teams assembled for roll call and the distribution of assignments. Each street cleaner wore an all-white uniform—jacket, shirt, pants, and pith helmet. Higher ranks wore gray. He asked for input from all workers. While focused on discipline, these measures seemed to elevate the worker (Nagle 2013, 104). In five weeks, they cleaned 113 miles of streets, as compared to twenty-six when Tammany was in charge ("Col. Waring's Methods" 1895).

Five Points was a notoriously dangerous neighborhood. In his typical confidence, Colonel Waring very publicly took up Five Points as "a test." He said, "and if we cannot clean that district, we might as well give up." He chose two men to lead this foray. One was a graduate of MIT. "The other," he reported, "was a young man from New Orleans, a Creole whom I picked because I liked his looks." He described a police escort and repeated trips into knee-high garbage. The crew gained the residents' trust ("Col. Waring's Methods" 1895). He turned street cleaners into heroes.

In six months, Colonel Waring took a *Times* reporter for a limitless inspection tour. Absent were the abandoned wagons, dead animals, stray paper, festering food, and ash. Everywhere, the skin of asphalt and stone streets were visible ("Clean Streets at Last" 1895).

The city, at Colonel Waring's urging, held parades, in 1896 and 1897, to honor the workers, attracting adoring crowds. The *New York Times* extolled, "The first parade of the street-cleaning force was the most cheering demonstration that has ever been made to the citizens of New York of the possibilities of their Municipal Government" ("Colonel Waring" 1898). By 1898, as the city was preparing for expansion, it was abandoning the funding of cleanliness. Colonel Waring tried to leave his successor prepared.

That year, following the U.S. victory in the Spanish-American War, President William McKinley, trying to protect U.S. troops, asked Colonel Waring to investigate Cuba's contagious epidemics. He spent two weeks, after which he returned to New York City to prepare his report. To his disbelief, he became acutely ill with yellow fever. He died October 29, 1898 ("Col. Geo. Waring Dead" 1898).

Underscoring his remarkable achievement with the workforce, the *New York Times* obituary began, "What more cruel stroke could there be of the irony of fate than the death of George Waring from a filth-disease?" ("Colonel Waring" 1898). Jacob Riis said, "It was Colonel Waring's broom that first let light into the slum" (Lee 2009). His understanding of contagion was wrong, but in a Gilded Age of corrupt, boastful power and caustic inequality, Colonel Waring's brief elevation of the work and worker was a cure. The city was, for the moment, clean.

References

Cassedy, James H. 1962. "The Flamboyant Colonel Waring: An Anti-Contagionist Holds the American Stage in the Age of Pasteur and Koch." *Bulletin of the History of Medicine.* 36(2): 163–176.

"Clean Streets at Last." 1895. *The New York Times.* https://timesmachine.nytimes.com/timesmachine/1895/07/28/106066139.html?pageNumber=28. Accessed on August 10, 2019.

"Col. Geo. E. Waring Dead." 1898. *The New York Times.* https://timesmachine.nytimes.com/timesmachine/1898/10/30/105965064.pdf. Accessed on August 20, 2019.

"Col. Waring's Methods." 1895. *Brooklyn Daily Eagle.* https://bklyn.newspapers.com/image/50404269/. Accessed on August 14, 2019.

"Colonel Waring." 1898. *The New York Times.* https://timesmachine.nytimes.com/timesmachine/1898/10/30/105965237.html?pageNumber=18. Accessed on August 10, 2019.

Lee, Jennifer S. 2009. "He Cleaned the Streets, and Left the Presidency to Others." *The New York Times.* https://

cityroom.blogs.nytimes.com/2009/10/01/he-cleaned-the-streets-and-left-the-presidency-to-others/. Accessed on August 20, 2019.

"A Man at the End of a Broom." 1895. *The New York Times*. https://timesmachine.nytimes.com/timesmachine/1895/03/30/issue.html. Accessed on August 20, 2019.

Nagle, Robin. 2013. *Picking Up: On the Streets and Behind the Trucks with the Sanitation Workers of New York City*. New York: Farrar, Straus and Giroux.

Jim Nordlinger is a freelance writer, specializing in medical science, education, and history, as well as an adjunct professor of education.

Composting Helps Keep the Soil Beneath Our Feet Healthy and Vibrantly Alive!

Lisa Perschke

One of the most important tools of waste management is composting. Many urbanites are stunned when they learn that the soil beneath their feet consists of a vast underground kingdom of microorganisms, without which, life as we know it would not exist. Healthy soil is alive with a complex array of macro- and microorganisms that constantly hunger for carbon. These tiny, diverse creatures make up our complex soil food web.

"A single teaspoon (1 g) of rich garden soil can hold up to 1 billion bacteria, several yards of fungal filaments, several thousand protozoa, and scores of nematodes," according to Kathy Merrifield, a retired nematologist at Oregon State University (Herring 2010). These tiny, diverse microscopic creatures are essential to the health of plants, trees, and other vegetation living within the soil's rhizosphere. This vital miniature ecosystem constantly hungers for organic materials. Composting invigorates the soil's food web community by providing the nutrients, moisture, and habitat needed by this vast array of life under our feet. By nurturing and maintaining our healthy, diverse soil biome, thousands of pathogens are kept in check.

"Special soil fungi, called mycorrhizal fungi, establish themselves in a symbiotic relationship with plant roots, providing plants not only the physical protections but also the nutrient delivery they need. In return for plant exudates, these microbes provide water, phosphorus and other necessary plant nutrients" (Lowenfels and Lewis 2006, 25). With mycorrhizal fungi present, healthy trees and plants can attract protective nematodes, other fungi, and bacteria to their root zone where they attack invading parasites. Bacteria, attracted to plant exudates on leaf surfaces, produce slime, which traps invading pathogens and pests. Fungi are miles and miles in length. By breaking down carbon materials at the soil's surface, fungi act as conveyor belts, transporting these important nutrients to tree roots below. Fungi assist trees and plants with communicating danger to other similar species in the region when a disease or pathogen has attacked. Fungi become root extensions, connecting these stationary species to one another. They aid trees and plants by transporting chemicals through the soil, so they can alert nearby species to become prepared, keeping such diseases from entering their systems. Healthy plants and trees also can send chemicals through the air through their stomata.

Composting is purposeful biodegradation of organic matter through an environmental process. Through composting, this diverse population of microorganisms utilizes organic materials, which they need to live, abundantly. The process requires several ingredients for it to work properly. Brown matter (e.g., autumn leaves, pine needles, and straw) and green matter (e.g., coffee grounds, grass clippings, and veggie scraps) are needed for their metabolic energy processes. Moisture is needed for organisms to breathe, eat, and move throughout the compost furrow structure. Healthy soil is needed to inoculate the new organic matter with trillions of diverse microorganisms. Oxygen is required to provide a healthy aerobic environment. These materials assist with turning and shredding by helping to ensure a quick, hot process that eradicates weed seeds and pathogens. This breakdown of matter slowly phases from a healthy, organic compost to a black, spongy, nutrient-rich material called humus.

The compost process recipe requires a 2:1 ratio of carbon to nitrogen. One part green matter, or nitrogen, is used for cell function and growth as well as enzyme and protein formation, which is used in the decaying process. Nitrogen attracts bacteria to this pile. Two parts brown matter, or carbon, is used for energy and building materials that cells need. Fungi hunger for carbon, and in exchange for this food source, these tiny strands possess the enzymes that are needed to break down the fibrous pile of matter. By ensuring that the compost pile is getting plenty of oxygen, composting involves turning and shredding of the organic matter. Such material movement encourages aerobic bacteria, such as psychrophilic, mesophilic, thermophilic and actinomycetes; microorganisms such as protozoa, rotifers, and nematodes; fungi; and other organisms to thrive in the compost and then in the soil.

Maintaining a well-balanced, moist organic pile ensures that the microbes can breathe, move, and eat the organic materials in the compost heap. Organic matter should be as moist as a "rung-out wet sponge." If compost becomes too soggy, the microbes will either drown and die or be washed away from the pile of materials, resulting in nutrient loss and an unhealthy anaerobic process, creating adverse materials and microbes.

Without oxygen in the composting process, materials will continue to decay, but anaerobically. Dangerous microorganisms and other chemicals such as alcohol, hydrogen sulfide, butyric acid, ammonia, vinegar, and lactic acid are produced, harming plants root cells even at minute amounts. Soils containing anaerobic microorganisms become depleted of iron and magnesium, which plants need. Compost made without oxygen can take up to one year to rid itself of these dangerous microorganisms; pH values in anaerobic compost are very acidic and need time to increase to a safer level of 6 or 7.

Anaerobically derived microorganisms in compost can cause appendicitis, diverticulitis, and perforation of the bowel in humans, when ingested. Anaerobic microorganisms destroy the beneficial aerobic microbe population.

Research has shown that rototilling destroys the structure of soil as well as reduces the biodiversity of the natural food web found there. Rototilling chops up the soil biota, breaks miles of established fungal hyphae, collapses millions of worm tunnels, and reduces the number of aerated pores existing between the soil particles. All physical actions occurring on the surface of the earth have a direct impact on the health of our soil. Even something that seems minimal, such as a quick walk across a garden plot, can compact our soil. When soil becomes compacted, it doesn't allow plant and tree roots the ability to expand and travel underground, causing various stem and root rot diseases. The compaction of the soil reduces the available oxygen levels, thus increasing the anaerobic bacteria population. Trees and plants become powerless over their environment and succumb to multiple diseases and other parasitic infestations in their world.

"Inorganic fertilizers, pesticides, herbicides, miticides and fungicides kill off the soil food web members and therefore have no role in composting" (Lowenfels and Lewis 2006, 124). Such chemicals can interfere with the composting process by either causing those beneficial "good" microbes to flee the area or making their bodies explode, which contributes to the heat and decay process that occurs in composting. These chemical products contain heavy salts and dangerous chemicals that are offensive to the bodies of bacteria and fungi. For soils that have received years of such chemical applications, utilizing a compost- or vermicompost-based tea that is completely aerated during its processing is very helpful. Such aerobic compost tea applications will inoculate such depleted soils and establish a healthy and diverse soil food web population. Cover crops, or green "living" mulch, can help to prevent weeds from germinating and add natural "needed nitrogen" to the soil.

By diverting food scraps and yard wastes that get mindlessly sent through our municipal trash cycle, we can minimize the amount of methane gas that we emit annually into the atmosphere. Organic wastes that are deposited in a landfill decompose anaerobically, with the absence of water and air, thus producing methane. Methane damages our environment twenty times more than carbon dioxide alone. Methane production, when trapped inside our atmosphere by our ozone layer, can increase our global temperature and cause climate change.

Composting puts organic waste where it is needed most, aerobically feeding the vast array of soil food web bacteria, microbial organisms, fungi, micro- and macroorganisms that are needed to maintain and nurture all living species flourishing in the soil beneath our feet.

References

Herring, Peg. 2010. "The Secret Life of Soil." OSU Extension Service. https://extension.oregonstate.edu/news/secret-life-soil. Accessed on August 20, 2019.

Lowenfels, Jeff, and Wayne Lewis. 2016. *Teaming with Microbes: A Gardeners Guide to the Soil Food Web*, 2nd ed. Portland, OR: Timber Press.

Lisa Perschke presently attends Eastern Michigan University's public administration master's Program. She works full time in Ann Arbor, Michigan, for the nonprofit organization, Recycle Ann Arbor. As recycle program coordinator, Lisa works with Ann Arbor businesses owners, property managers, custodians, maintenance staff, and residents, teaching ways to reduce consumption and increase recycling diversion rates through best management practices.

The Flow of Waste: Earth Day to the Tesla Gigafactory

Nidia K. Trejo

In the late 1800s, waste management was big news, as sanitation systems reached urban communities like New York City.

During World War II, recycling initiatives emerged, with the government encouraging collection of scrap metal, paper, and rubber—very valuable raw materials for the time.

By the 1970s, the U.S. Environmental Protection Agency stepped up recycling campaigns, supported the very first Earth Day, and started identifying hazardous waste sites in the 2000s. For decades, China welcomed U.S. waste streams. However, now the United States is back to managing its own waste stream.

The Nature of Nonhazardous Waste

Nonhazardous waste includes pretty much everything that can go into trash and recycling bins. It doesn't pose direct harm to humans and the environment. After trash pickup, the final destination is a landfill or an incinerator. Currently, there are over 2,500 U.S. landfills, with 24 percent actively accepting trash.

Call for Clean Trash

Concerns raised about landfills are that they contribute to greenhouse gas emissions and that they take up valuable space. With all the stuff U.S. consumers buy and throw away on a daily basis, it seems like trash will just keep piling up.

To deal with these concerns, the top three solutions looking into 2020 include:

1. Use the gas of landfills: Landfills generate methane, a greenhouse gas. The United States is promoting innovative projects to capture the methane and generate energy from landfills.
2. Build a city: Building on top of landfills is one solution to the space problem. It opens up hundreds of acres of land. Battery Park City in New York City and the Marina District in San Francisco are a few examples.
3. Remove the food: States like California found that food waste makes up to 40 percent of the total trash volume.

Diversion projects are underway mainly by creating home composting gardens and increasing donations to food banks.

What is the scale of impact? These three diverse ideas can potentially trickle into all fifty states and U.S. territories as solutions to common landfill concerns.

Recycle and Phase Out

Once recyclable materials are picked up from a home or business, they arrive at a sorting facility. Paper, cardboard, wood, and metal are recycled in the United States. However, difficulties arise with plastics.

The recycling plastic challenge is thrown into the hands of other countries or the world's oceans. With the Chinese import ban of 2018, the United States no longer exports hard-to-recycle materials to China. In the interim, plastics enter U.S. landfills and incinerators while better U.S. recycling infrastructure develops.

The top three solutions looking into 2020 include:

1. Recycle plastic for fashion: Unifi uses plastic water bottles as a raw material to produce polyester fibers. Some sportswear apparel brands, like The North Face and Adidas, love it. It is a second-life opportunity for the plastic waste stream to enter the fashion industry.
2. Artificially intelligent robots: AMP Robotics deploys automated robots to simplify the sorting process. Computer vision and machine learning algorithms identify materials to properly categorize into recycling. They pick up on details like color, shape, and size to help increase recycling rates across cities.
3. Eliminate plastic waste: Considering that 91 percent of plastic isn't recycled, cities and food brands respond by banning single-use plastic, like straws. The plastic straws

are quickly becoming an item of the past in cities like Seattle and Los Angeles. Nestlé and Starbucks are brainstorming alternatives to single-use plastic.

What is the scale of its impact? Plastic waste is a growing responsibility of the product developer and the original manufacturer. This leads global brands to rethink product life cycles in a circular perspective, not linearly.

The Nature of Hazardous Waste

Hazardous wastes are materials that can potentially harm human health and the environment. It can catch fire easily, burn easily, explode, be reactive, or destructive in a corrosive way. Think of batteries and used automotive oil.

Where is the hazardous waste going?

Hazardous waste is tightly regulated by the U.S. Department of Transportation. It enters transfer stations and then goes to North American landfills or overseas. Enforcing engineering controls is the traditional approach to safety. Making hazardous waste nonexistent is also an option.

The top three solutions looking into 2020 include:

1. Engineering controls: Crude oil refineries rely on automated sensor controls to contain their raw material and reduce risks of accidental spills. Converting oil into petroleum is an intensive chemical and engineering process. In an effort to close the loop, Veolia recycles residual oil from refineries.
2. Battery recycling: Although lead-acid batteries in automobiles are only considered hazardous if they spill, the use of lithium-ion batteries in automobiles can reduce hazardous waste. Tesla's Gigafactory is focusing on recovering valuable metals from lithium-ion batteries at end-of-life as well as scrap from the manufacturing process.

3. Green chemistry adoption: Pharmaceutical and hydraulic fracturing industries welcome safer alternatives at the building block, chemical stage. The green chemistry approach is to design chemicals upstream to reduce or eliminate hazardous qualities downstream.

What is the scale of its impact? Hazardous waste is bound to exist across many industries and products we use every day. It is a continuous balance between the best available materials to meet human needs and technological advances that allow for improvements.

Nidia Trejo earned her bachelor's and master's degrees of science from the University of California Davis and Cornell University, respectively. She is a technical writer, covering innovation in the fields of science, engineering, and technology. She carries international and interdisciplinary perspectives from her work into the global textile industry.

Waste Management and Tribes

Lynn Zender

There are 573 federally recognized tribes in the United States, and each has a unique waste management profile and corresponding issues. However, there are overarching issues that present unique challenges to tribes in providing waste management programs for their communities that are protective, financially feasible, and logistically practicable.

Legal issues

Tribes are legally sovereign nations, dependent on the federal government for certain services through what is termed "trust" responsibility. Like states, they are able to operate most federally delegated programs, including programs under every major environmental statute with the exception of one—RCRA, with which they are relegated to "treatment as a municipality." To operate a permitted landfill, they must apply to the state.

Why is this a problem? Throughout history, Supreme Court decisions and Congressional plenary power have waxed and waned in recognizing the extent of tribal sovereignty. This sovereignty essentially documents that tribes persist as a separate cohesive people in the face of government-sanctioned policies that have been universally recognized to have inflicted trauma and injustice.

While it may make financial and logistical sense for a tribe to operate a landfill, applying to a state's RCRA Subtitle D program is the mechanism by which that state gains enforcement authority over the landfill. And many tribes and legal experts view the act of acceding to state jurisdiction as risking the erosion of legally recognized tribal sovereignty. With the exception of Alaska, most tribes have county boundaries dissecting their reservations, and non-Indian-owned "fee lands" within their boundaries. Other lands both on and off the reservation are disputed between tribes and states. A tribe's sovereignty can extend over non-Indians and fee land if there is serious impact to the health and welfare of the tribe. But the Supreme Court has found that a tribe can implicitly relinquish its zoning authority over fee lands simply by participating in a county hearing process to raise their objections. Applying for a state permit may not be an act that tribes are willing to do.

Imagine a county or city government without the recourse to fines and laws regarding community waste disposal. Some tribes and county/state governments have reciprocal relationships, such as cross-deputization, and work well with each other in enforcing laws within the reservations, while others do not. One area where this is a problem is unauthorized dumping. Tribes do not have criminal, nor in some cases civil, authority over most non-Indians and can be hard-pressed to exert their authority when it comes to waste. But state highway or sheriff patrol might be reluctant to cite illegal dumpers and litterers on fee land or right-of-way highways even when they have the authority to do so because of the jurisdictional problems it can exacerbate. Open dumping of project waste by contractors can be an issue, and tribal nonmembers can use tribally funded waste transfer stations without contributing to their operations. Tribes are hard-pressed to cite nonmembers who can refuse to appear before tribal courts or pay tribal fines. If nonmembers bring a case to the state court, it may be decided against the tribe, endangering its recognized jurisdiction.

Infrastructure issues

Most tribes are small, rural, and remote. Economies of scale dictate that waste management will be costly relative to the number of households and area covered. And low-income status and poverty is prevalent. In 2016, single-race American Indians and Alaska Natives had median household incomes that were only 68.9 percent that of the national average, and 26.2 percent lived in poverty, the highest rate of any race group. So tribes often must choose between offering supplementary waste services, such as recycling centers and household hazardous waste collection, by coordinating with the county, or doing without. The ability to do so depends on the political relationship between the county/state and the tribe and on what services the county offers. Rural counties often are low income as well and may not offer supplementary waste services. As a result, a tribal "transfer station" can be an unlocked and unstaffed dumpster on the side of a road, lending itself to abuse by tribal members and nonmembers alike.

The 229 Alaska tribes provide an extreme example of substandard infrastructure. The State of Alaska's singular program permits landfills for small and remote communities, but which are unlined, infrequently covered, unvented, and without leachate treatment. Contained community-scale open burning in steel drums or cages is allowed and widely practiced. Because these communities are inaccessible by road and are isolated from each other, use of regional facilities is not possible. Communities receive only by bimonthly summer barge and small passenger plane service. Each community of fifty to 1,000 residents operates a landfill that is managed only up to the state's minimal requirements.

Due to the extreme cost of road construction, about three-quarters of these landfills are within one mile of homes. With no control over waste burn unit emissions, toxic smoke can be smelled and breathed by residents regularly in a number of communities. Hazardous wastes, electronics, and other wastes with toxic metals and contaminants are by and large discarded at the unlined landfills. The communities have largely traditional ways of life, including hunting, fishing, and gathering to supply much of their dietary needs. Water is also locally sourced. The dependence on local water and food is important because the landfills are proximate and typically hydrologically connected to rivers and surface waters that support these resources.

These households average under 35 percent of the state's average median household income. Their budgets are stretched to cover heat, fuel, and gas for subsistence activities. Paying for even basic infrastructure like heavy equipment, fencing, and storage facility is difficult. Operator hours are consequently minimal even while tasked with substantially greater challenges than their urban counterparts.

Waste management here is on the frontline of public health. Babies born to mothers residing in Alaska villages with landfills ranked of greatest exposure potential, and having high hazard contents (e.g., batteries) presented increased risks for lower birth weight and shorter pregnancies; for landfills with high hazard contents, there is a 4.3 times greater risk for certain non–life threatening birth defects. Some tribes are able to operate backhaul programs to reduce hazardous waste in their landfills, but for the majority, sustained and comprehensive programs to do so are infeasible due to the high cost of shipping to Seattle.

Exposure risk issues

Product waste disposal and transfer sites are a main means by which toxic chemicals make their way into the natural environment. And tribes' customary and traditional practices center in the natural environment. Consequently, tribes tend to be more exposed, and in different ways, to the contaminants that they and neighboring communities manage. Their water and food sources may be directly at risk from leachate migration, flooding, overland flow, or particulate settlement from open burning. The tribes' stake in devising a protective waste management strategy is very different than for conventional urban communities whose populations have very little contact with waste or its leachate and emissions.

The tribes' challenges in reducing their people's exposure to chemicals in waste is compounded by a federal law called the Toxic Substances Control Act (TSCA). TSCA regulates the use of chemicals in consumer products—from toys, furniture, and frying pans to computers, glue, and paper. It mandates the evaluation of a chemical through its entire life cycle, including product disposal. The exposure risk must be found to be reasonable, or a chemical risk management plan must be developed.

But because TSCA is a national-level law, it is that conventional urban community exposure to chemicals from waste disposal that is evaluated, and hence, it is that general population that is protected in determining chemical restrictions needed to protect the public from toxic chemical release from waste disposal practices. Tribes cannot presume that what may otherwise be nominally responsible management practices for a conventional community will protect their members from unreasonable chemical risk emanating from the waste they are managing. They must be cognizant of the special relationship their members have with the environment and the various ways they may be exposed in carrying out customary and traditional ways of life.

Dr. Zender is director of Zender Environmental Health and Research Group, an Anchorage nonprofit that conducts research, training, and policy development on tribal waste management issues in Alaska and nationally, www.zendergroup.org.

Sludge: The Good, the Bad, and the Ugly
Geena Zick

Sludge, also known as biosolids, is the leftover product at the bottom of a sewage tank after the treatment of wastewater. It is human feces and a mix of other things that were flushed down toilets and dumped down drains.

The Good

To be frank, there is nothing good about sludge. It just made a catchy title.

The Bad

Sludge contains everything that is flushed down toilets or dumped down drains. Because of this, it has been found to contain heavy metals including chromium, mercury, and lead; pathogens; pharmaceuticals; dioxins; and per- and polyfluoroalkyl substances (PFAS). These are known carcinogens and neurotoxins. In 2003, the EPA announced its decision not to regulate dioxins in land-applied biosolids (U.S. Environmental Protection Agency 2019a).

Most people assume that when they flush something down the toilet or dump something into a drain, they will never see it again. Unfortunately, this is not always the case. When sludge is applied to farmland, these contaminants are absorbed by crops (Shinbrot 2013). People then unknowingly consume these contaminants. We can do more to keep these toxic chemicals out of our waste.

The Ugly

Sludge used to be dumped into the ocean (U.S. Environmental Protection Agency 1992). In 1992, this was outlawed, so the EPA turned to the land. Sludge is often referred to as "organic" or "nutrient rich," making it seem good for our environment (U.S. Environmental Protection Agency 2017). It is claimed that land-applied sludge is regulated and that pollutant levels are too low to cause harm to humans (U.S. Environmental Protection Agency 2019b). However, not all pollutants are regulated or even known, and regulated levels are often not health protective.

The toxic chemicals known as PFAS were created in the 1930s. PFAS are in nonstick cookware, stain-resistant clothing, and some firefighting foams also contain PFAS, which is why many military bases are highly contaminated. PFAS are carcinogens and can cause kidney problems, reproductive disorders, and birth defects (Agency for Toxic Substances and Disease Registry 2019). These substances are "forever chemicals" because they are extremely resilient and do not break down in the environment or in our bodies. Almost every person has low levels of PFAS in his or her body (Agency for Toxic Substances and Disease Registry 2019). Because of this resiliency and common use, human waste often contains PFAS.

PFAS has been found in land-applied sludge in multiple states. This year, Maine decided to test their sludge and found all of it contained PFAS (Lerner 2019). Although the state knows it contains a dangerous contaminant, they agreed to continue to spread the sludge on farmland. The state is requiring farmers to test their fields for contamination before applying the PFAS sludge. If the field is below a set standard of PFAS, they calculate how much the contamination would increase with the application of PFAS-contaminated sludge, and if that is a low enough level, the sludge is applied. This encourages sludge application on unpolluted, clean fields and perpetuates the myth that "dilution is the solution to pollution."

In 2002, Dr. David Lewis, a top scientist for the EPA, published a paper citing evidence that sewage sludge applied to land had killed a local resident. This man's name was Shayne Conner, and his illness puzzled doctors. A similar death occurred the year before to a young boy named Tony Behun. They both had respiratory problems, skin lesions, and a fever caused by an infection from *Staphylococcus aureus* (Lewis et al. 2002).

Lewis researched the dangerous health impacts of sludge and was forced out of the EPA. Along with two Georgia farmers, he sued the EPA for publishing false data to support using sludge as a fertilizer (Tollefson 2008). The lawsuit claims that the study concealed evidence of sludge leading to cattle deaths on two farms. The U.S. National Academy of Sciences (NAS) published a report in 2002 citing the study, which stated that livestock deaths have not been linked to sludge. The NAS report also recommended the EPA survey sludge for chemicals, track health complaints, and conduct studies to assess the impacts of biosolids (Tollefson 2008). It took the EPA six years to begin this work.

The Solution

The real solution is for companies to stop the manufacture and use of toxic chemicals. Most people are not thinking about toxins in our food, body, land, and so on every day, and that needs to change. As a society, we need to be conscious of what we're putting into our bodies and what is being put out into the environment. If our sludge did not contain these toxins, it would be a great fertilizer for our crops, like organic and hormone-free and antibiotic-free animal manure. Our natural reaction is to spread sludge on land, and that would be OK if sludge was still "natural."

In the meantime, we should ban the land-application of sludge, develop programs to reduce the cost of landfill disposal of sludge, require long-term maintenance of sludge sites, and enact a strong "toxic-use reduction" program. These are short-term solutions and are just the beginning of the battle with sludge.

References

Agency for Toxic Substances and Disease Registry. 2019. "Per- and Polyfluoroalkyl Substances (PFAS) and Your Health."

https://www.atsdr.cdc.gov/pfas/index.html. Accessed on August 24, 2019.

Lerner, Sharon. 2019. "Toxic PFAS Chemicals Found in Maine Farms Fertilized with Sewage Sludge." *The Intercept.* https://theintercept.com/2019/06/07/pfas-chemicals-maine-sludge/. Accessed on August 19, 2019.

Lewis, David L., et al. 2002. "Interactions of Pathogens and Irritant Chemicals in Land-Applied Sewage Sludges (Biosolids)." *BMC Public Health.* 2(11): 409–423.

Shinbrot, Xoco. 2013. "Biosolids or Biohazards?" *Pesticides and You.* 32(3): 10–11.

Tollefson, Jeff. 2008. "Raking through Sludge Exposes a Stink." *Nature.* 453: 262–263.

U.S. Environmental Protection Agency. 1992. "EPA Declares End of Ocean Dumping as New York City Ceases; EPA Committed to Long-Term Beneficial Alternatives." https://archive.epa.gov/epa/aboutepa/reilly-new-york-mark-end-sewage-sludge-dumping.html. Accessed on August 24, 2019.

U.S. Environmental Protection Agency. 2017. "Basic Information about Biosolids." https://www.epa.gov/biosolids/basic-information-about-biosolids. Accessed on August 24, 2019.

U.S. Environmental Protection Agency. 2019a. "Dioxins in Sewage Sludge." https://www.epa.gov/biosolids/dioxins-sewage-sludge. Accessed on August 19, 2019.

U.S. Environmental Protection Agency. 2019b. "Frequent Questions about Biosolids." https://www.epa.gov/bio solids/frequent-questions-about-biosolids. Accessed on August 24, 2019.

Geena Zick is a rising senior studying wildlife biology and anthropology at the University of Vermont.

This story of waste management is, to a large extent, the story of individuals and organizations who have believed in, worked for, and promoted the topic for much of their existence. The list of such individuals and organizations is very long. The entities selected for this chapter can do no more than represent the work of countless other men, women, and groups struggling to understand how best to deal with the municipal, industrial, agricultural, mining, and other wastes that are an integral part of human civilization.

Air & Waste Management Association

The problem of urban smoke is hardly a new one in human history. That problem reached a peak in the United States, however, in the early 20th century. During the decade between 1900 and 1910, many major cities, including Chicago, Cincinnati, Cleveland, Denver, Detroit, Indianapolis, Milwaukee, Philadelphia, Rochester, St. Louis, St. Paul, Syracuse, and Toronto had passed smoke abatement ordinances to deal with the problem. In 1907, a group of municipal smoke inspectors interested in the specific issue of the connection between smoke and human health issues met in Milwaukee to discuss issues of common interest. Participants at the meeting decided

Edwin Chadwick was a British social reformer who devoted his life to sanitary reform as Britain was becoming increasingly urbanized and industrialized in the 19th century. (National Library of Medicine)

to create a new organization—the Air & Waste Management Association (A&WMA)—to pursue these issues further. That organization remains in existence today, with more than 5,000 members in sixty-five countries.

Those members are divided into thirty-four sections and sixty-five chapters worldwide, where members meet and work with close associates on issues of local or regional importance. Among these groups are the East Central Section (East Michigan, Indiana, Kentucky, Northern Ohio, and Southwest Ohio chapters), the Southern Section (Alabama, Eastern Tennessee, Georgia, Mississippi, and Western and Middle Tennessee chapters), and West Coast Section (Channel Islands, Delhi, Mid-Pacific, Mojave Desert, Orange County, San Diego, Thailand, and Singapore chapters). A general overview as well as a detailed list of instructions and information about A&WMA itself, its sections, and its chapters is included in a manual called *Manual of Operations: Sections & Chapters Council.*

A&WMA sponsors, cosponsors, and participates in several national and international meetings, such as the IUAPPA (International Union of Air Pollution and Environmental Protection Associations) World Clean Air Conference; International Conference on Thermal Treatment Technologies and Hazardous Waste Combustors; and the A&WMA conferences on Freight & Environment: Ports of Entry Conference and Atmospheric Optics: Aerosols, Visibility, and the Radiative Balance. The organization also sponsors an annual conference as well as sectional meetings. A&WMA also offers periodic webinars on topics of interest to members. A recent webinar, for example, dealt with "Living with PFAS (perfluorinated alkyl substances)—Managing the Investigation, Waste, and Remediation Challenges."

A&WMA also sponsors two important publications: (1) a weekly online newsletter called *A&WMA Newswire*—which keeps members up to date with recent developments in the field, (2) a monthly magazine, *EM*—which provides information on regulatory changes; EPA research; new technologies;

market analyses; environment, health, and safety issues; new products; and professional development opportunities.

Robert Bullard (1946–)

Robert Bullard is widely known as the father of environmental justice. Several of his books have become classics in the field, including *The Quest for Environmental Justice: Human Rights and the Politics of Pollution* (2005); *Growing Smarter: Achieving Livable Communities, Environmental Justice, and Regional Equity* (2007); *The Black Metropolis in the Twenty-First Century: Race, Power, and the Politics of Place* (2007); *Environmental Health and Racial Equity in the United States: Strategies for Building Environmentally Just, Sustainable, and Livable Communities* (2011, with Glenn S. Johnson and Angel O. Torres); and *The Wrong Complexion for Protection: How the Government Response to Disaster Endangers African American Communities* (2012, with Beverly Wright). Bullard's book, *Dumping in Dixie* (1990, 2000, 2004), is regarded by some authorities as one of the most powerful statements on the issue of environmental justice yet to be written.

Robert Doyle Bullard was born in Elba, Alabama, on December 21, 1946. He received his bachelor's degree in government from Alabama A&M University in 1968, his MA in sociology from Atlanta University in 1972, and his PhD in the same field from Iowa State University in 1976. He then worked as an urban planner in Des Moines, Iowa (1971–1974), administrative assistant in the Office of Minority Affairs at Iowa State (1974–1975), research coordinator in Polk County, Iowa, and director of research at the Urban Research Center of Texas Southern University (1976–1978).

In 1976, Bullard was also appointed assistant professor at Texas Southern, and in 1980, he was promoted to associate professor. He then went on to hold a series of academic appointments at Rice University (1980), the University of Tennessee (1987–1988), the University of California at Berkeley

(1988–1989), the University of California at Riverside (1989–1993), and the University of California at Los Angeles (1993–1994). In 1994, Bullard was appointed Ware Distinguished Professor of Sociology and director of the Environmental Justice Resource Center, Clark Atlanta University, a post he held until 2011. He then accepted an appointment as dean of the Barbara Jordan-Mickey Leland School of Public Affairs at Texas Southern University. Five years later, he was appointed to his current position as Distinguished Professor of Urban Planning and Environmental Policy at Texas Southern.

Bullard has received a host of awards and honors, including two from the Gustavus Myers Center for the Study of Human Rights in North America for his books: the first for *In Search of the New South: the Black Urban Experience in the 1970s and 1980s* in 1980 and the second for *Residential Apartheid: the American Legacy* (edited with J. Eugene Grigsby III and Charles Lee) in 1996. His other awards include the William Foote Whyte Distinguished Career Award of the American Sociological Association's Section on Sociological Practice (2007); Co-op America, Building Economic Alternative Award (2008); Planet Harmony, "Ten Green Heroes" Award (2010); The Gio, "100 Black History Makers in the Making" (2010); American Bar Association award for Excellence in Environmental, Energy, and Resources Stewardship (2015); and Children Environmental Health Network (CEHN), Child Health Advocate Award (2017). In 2014, the Sierra Club created a new Environmental Justice Award in his name. In addition to the eighteen books he has authored and edited, Bullard has been the author of more than seventy-five peer-reviewed papers in scholarly journals and book chapters.

Dollie Burwell (1948–)

Dollie Burwell was a leader in the protests in Warren County, North Carolina, in the late 1970s and early 1980s against the dumping of 32,000 cubic yards of soil contaminated with polychlorinated biphenyls (PCBs). When that effort was unsuccessful, she later spearheaded efforts to clean up the dump area and restore it to its natural condition, a campaign that eventually was successful. Nearly fifteen years after the initial dumping in Warren County, the state of North Carolina allocated $1 million to study the toxicity of soils at the Warren County dump. As a result of that study and continued pressure by residents of the county, the state then allocated an additional $15 million for cleanup of the dump. That task was finally completed, and the former dump has become clean enough that it is now used as a recreational site.

Dollie Burwell was born in Warren County on May 24, 1948. She became interested in politics early in life and worked on a voter registration project at the age of twelve. Although she came from a very poor family, she was determined to have a college education and took courses at Durham College and Shaw University. In 1971, she married William J. Burwell and moved with him to an army post at West Point. There she continued her political activities by organizing couples of color who had been denied government housing because of their race. The Burwells eventually returned to Warren County, where Dollie became involved in the dumping protest, while also attempting to raise her two young daughters. The older of those daughters, Kimberly, has also become an activist in the environmental justice movement.

Over the four-decade battle in Warren County, Burwell continued to work on issues of environmental justice and other human and civil rights problems. In addition to her private efforts, she has been involved as aide to Congresswoman Eva Clayton (D-NC; 1992–2003), and later to Congressman G. K. Butterfield Jr. (D-NC; 2004–). In these posts, she has served as a link between the Congress member and his or her constituents, has represented the congressperson at various events within the district, has been a caseworker in three towns in the district, and has served as the congressperson's field representative in the towns of Rocky Mount, Granville, Halifax, Person, and Vance.

Edwin Chadwick (1800–1890)

Chadwick is perhaps best known today for a study he conducted from 1839 to 1842 on the health conditions of the general population of England. He paid special attention to the living conditions of poor people in the city and how those conditions might contribute to disease and other health disorders among that population. He was assisted in his research by three physicians, Neil Arnot, Thomas Southwood Smith, and James Kay-Shuttleworth. The study was historic in that it was the first time in British history that medical men chose to investigate the lives and health conditions of a large population of average men and women in the country. The group's report, "Report on the Sanitary Condition of the Labouring Population of Great Britain," is generally regarded as the beginning of the nation's new interest in public health, leading eventually to the first major legislation in that area, the Public Health Act of 1848. Although somewhat ineffective in its first incarnation, the act eventually became the model on which additional public health laws through the end of the century were based.

Edwin Chadwick was born in the neighborhood of Longsight in the city of Manchester on January 24, 1800. His father was James Chadwick, a well-known and highly regarded liberal politician who, among his other accomplishments, is remembered as a tutor to young John Dalton. Edwin received his early education at local schools in Longsite and the Manchester suburb of Stockport. When his family moved to London in 1810, his education continued with private tutors. He eventually decided to study law and was admitted to the Inner Temple in 1830. His law career was insufficient, however, to provide him an appropriate level of living, and he began to take on other jobs. One of these jobs was as a freelance writer, a position in which he was able to express his views on a host of social, economic, political, and other issues of the time. His ideas grew, to large extent, out of his connections with and admiration for the most famous social reformer of the time, Jeremy Bentham.

His first formal job came in 1832 when he was appointed a member of the newly created Royal Commission into the Operation of the Poor Laws. The commission was established to collect data on the existing system for meeting the needs of the poorest people in the nation and to make recommendations for necessary changes in the system. The commission's final report, issued in 1833, made the revolutionary suggestion that the nation's poor laws be administered on a national level, not, as was then the practice, by local units with very different philosophies and modes of action. Within a year's time, the commission's recommendations had been adopted, largely in toto, in the form of the Poor Law Amendment Act of 1834. In the same year, Chadwick was appointed secretary of the Poor Law Commission, created to carry out the provisions of the 1834 law.

When the first Board of Health was created in London in 1848, Chadwick was appointed as one of its members. He held that post for six years until he was "pensioned off," apparently because of his abrasive approach to problems involving sanitation. He generally held strong opinions as to necessary changes in order to improve public health and was not always diplomatic in his expression of those views. Although his removal from the Board of Health marked the end of his formal career in the field, he continued to write and speak about reform politics and remained active in the field. For example, he pressured Home Secretary Lord Palmerston during the Crimean War (1854–1856) to create a commission to study health conditions of troops in the field, inquiring into and relieving the sufferings of the troops. His suggestion was responsible, in part, for creation of the Administrative Reform Association in 1855 to further study the issue of military health. Chadwick was especially active in the group, although his contributions appear to have been less than one might have hoped because he "was a bad pamphleteer—his arrogant style merely increased opposition." (Bloy, Marjie. 2002. "Edwin Chadwick (1800–1890)." The Victorian Web. http://www.victorianweb.org/history/chad1.html. Accessed on July 30, 2019.)

In 1858 he raised the subject of defective sanitation in the Indian army: the support his views gained afterward led to the appointment of the Indian army sanitary commission.

Chadwick's contribution to public sanitation and a host of other public issues were largely ignored by official agencies until 1889, when he was knighted and made Sir Edwin Chadwick. He had by that time also been elected to several foreign scientific, medical, and other societies, including those of Belgium, France, and Italy. He died on July 6, 1890, at his home at Park Cottage, East Sheen, Surrey.

Environmental Protection Agency

The Environmental Protection Agency (EPA) is the nation's primary agency for carrying out and disseminating research on the environment as well as writing rules and regulations and carrying out legislative actions designed to protect and preserve the environment. The agency was created under provisions of the Reorganization Plan No. 3 in the administration of President Richard M. Nixon on July 9, 1970. The agency began operation five months later, on December 2, 1970. At the outset, the agency was regarded largely as a technical department, responsible for setting goals and standards for environmental policy. Over the years, its purview has been vastly expanded by several acts of legislation. When first established, the EPA had about 5,800 employees and a budget of $1.4 billion. As of fiscal 2019, those numbers had increased to 14,172 employees and a budget of $8,849,488,000. The EPA carries out many of its functions through ten regional offices in Boston, New York City, Philadelphia, Atlanta, Chicago, Dallas, Kansas City, Denver, San Francisco, and Seattle. The San Francisco office serves in addition to four western states, the territories of American Samoa, Commonwealth of the Northern Mariana Islands, Federated States of Micronesia, Guam, Marshall Islands, and Republic of Palau. The Seattle office includes in its area of responsibility 271 Native American tribes.

The primary areas for which EPA is responsible are air; chemicals and toxics; greener living; health; land, waste, and cleanup; science; and water. Waste management activities fall primarily within the areas of greener living (food waste and recycling); land, waste, and cleanup (landfills, hazardous waste, plastic and waterways, superfund, and cleanups); and water (wastewater, storm water, and runoff). Some of the issues of specific concern at the end of the 2010s are garbage (solid waste, including landfills, food waste and recovery, electronic waste [e-waste], and the Resource Conservation and Recovery Act); cleaning up land and emergencies (cleanups, cleanup locations in your community, land revitalization, brownfields, superfund cleanup program, renewable energy sites on contaminated land, and emergency response); hazardous waste (hazardous waste test methods; listed, characteristic, and mixed radiological wastes; and e-manifest and other resources for hazardous waste handlers); and reduce waste (plastic- and trash-free waters, recycling, composting at home, sustainable materials management, and waste generation facts and figures). With its vast collection of statistics and data, guidelines, reports, publications, and other features, the EPA website is the first place to go, in most cases, on almost any topic in the field of waste management.

The EPA conducts its own original research, provides grants to independent researchers, and draws on the results of research in many fields of environmental protection in order to carry out its basic mission. The Office of Research and Development maintains seven research laboratories dealing with specific topics: computational toxicology, environmental assessment, environmental research, exposure research, health and environmental effects, homeland security, and risk management. Other research projects, carried out at other EPA laboratories, focus on topics such as air, climate, and energy; chemical safety for sustainability; human health risk assessment; homeland security; safe and sustainable water resources; and sustainable and healthy communities. The agency also draws on the expertise of specialists in most of fields of research. Some examples are the Advisory Council on Clean Air Compliance, Board of Scientific Counselors, Clean Air Scientific Advisory Committee, National Advisory Council for Environmental Policy and Technology, and Science Advisory Board. Three EPA offices focus their research and studies on specific geographical areas and problems. They are the Chesapeake Bay Program Office, Great Lakes National Program Office, and EPA Region 4 Gulf of Mexico Division.

The EPA has long been, and probably always will be, an agency embedded in political controversy. It is at the center of an endless crucial debate over protection of the natural environment versus economic development. As an example, climate change has become one of the most contentious issues in American politics in the 21st century. Some individuals argue that the very survival of human civilization is threatened by a changing climate. These individuals argue that aggressive actions must be taken as soon as possible to deal with climate change. For these individuals, climate change should be a critical, if not the single most important, topic in the agency's agenda. Any presidential administration that shares that view is likely to increase funding and other forms of support for the work of EPA.

Other individuals may agree that climate change is an issue that deserves our attention. However, the economic costs of making excessive changes in our environmental policy are too great to focus very heavily on environmental concerns. Such individuals may argue, for example, that removing limitations on the use of fossil fuels causes more disruption than can be justified by any possible climate change scenario. Any presidential administration that adopts this view, then, is likely to cut back on EPA funding and on guidelines and regulations that sacrifice economic development in the name of environmental responsibility.

Alfred Fryer (1830/31–1892)

Alfred Fryer's place in waste management history is due to his invention of the first waste incinerator, known at the time as a *destructor*. He was born in Rastrick, Yorkshire, in either 1830 or 1831 (sources differ). The first we know of his life comes in 1865 from the invention of his device known as the *concretor*. It was developed as a way of improving the method by which sugar was transported. At the time, sugar was transported as the juice of the sugarcane plant, a process that resulted in substantial amounts of loss. Fryer's concretor converted the liquid sugarcane to a solid form, thus making it more easily transportable with less product loss.

At the time, Fryer was a partner in the firm of Fryer, Benson, and Forster, a sugar refinery with 170 employees. He came to that business after spending the early part of his life as a dealer in teas. In 1874, Fryer applied for a patent for his destructor, the first known device designed for the incineration of waste materials. The ash produced by this process was not thrown out, but was mixed with lime to make bricks or with cement to form paving blocks. The destructor was, therefore, both the first waste incinerator and an early example of the recycling of waste products. Fryer also established his own company, Fryer Concrete, to take advantage of one of his destructor system's by-products, the concrete formed in the last stages of operation. That company lasted until 1885.

Fryer retired from active involvement in the business world in 1884 to his home in Wilmslow. He did not retire from an interest in the world around him; one of his first projects was the construction of an astronomical observatory at his home. He also developed friendships with several important scientists of the time, including Norman Lockyer, Richard Proctor, and Balfour Stewart. He also devoted his time to the writing of a number of books, including *Peculiarities of Vital Statistics of the Society of Friends*, *Floating Lightships*, *Influence of Forests on Rainfall*, *Balance of Trade*, *Cost of Living in Various Countries*, *The Silver Question*, and *Wilmslow Graves—a Local History*. He also edited an American text, *American Investments* for the publisher, Jarvis Conklin Co., in Kansas.

Fryer died at his home in Wilmslow on December 13, 1892.

Global Alliance for Incinerator Alternatives

Global Alliance for Incinerator Alternatives (GAIA) is a worldwide alliance of more than 800 grassroots organizations, nongovernmental groups, and individuals from more than ninety countries who are interested in learning more about and educating the general public about methods of waste management that do not include incineration. It was founded at a meeting in December 2000 in South Africa of eighty people from twenty-three countries. Some of the organizational members of GAIA are (in Canada and the United States) the Toxic Action Center, Minnesota Public Interest Research Group, Prevent Cancer Now, Eco-Cycle, and Zero Waste San Diego; (in Latin America) BIOS Argentina, Fundación el Árbol, Ciudad Saludable, Aliança Resíduo Zero Brasil, SiKanda, and Fundación Basura; (in Europe) Amigos de la Tierra (Spain), Zero Waste France, The HuMuSz—Waste Reduction Alliance (Hungary), Zelena Akcija (Croatia), and Aksjon Steng Giftfabrikken (Norway); (in the Middle East and Africa) IndyACT (Lebanon), Green Africa Youth Organisation (Ghana), Development Indian Ocean Network (Mauritius), Surfrider Foundation Maroc (Morocco), and Zéro Déchet Sénégal; (Asia and Oceania) Shikshantar (Rajasthan, India), Perkumpulan Gerakan Indonesia Diet Kantong Plastik (Indonesia), Green Crusaders (Malaysia), and Zero Waste Kamikatsu (Japan).

GAIA has claimed a number of accomplishments in its two-decade history. These include:

- Stopping the transfer of hazardous incinerator ash from the United States to El Salvador (2000)
- Organizing the first "Waste Not" meeting in Bangkok (2000)

- Working toward the ban of incinerators in the state of Delaware (2000)
- Shutting down the Aloes Medical Waste Incinerator in South Africa (2001)
- Launching the first global campaign to stop the World Bank from funding incinerators (2001)
- Contributing to Bangladesh's ban on plastic bags, the first in the world (2002)
- Publication of "Waste Incineration: A Dying Technology" (2003)
- Organizing the first GAIA meeting in the United States (Hartford, Connecticut, 2006)
- Releasing the film *Story of Stuff* (2007)
- Organizing the first "waste-to-energy" meeting in India (Kerala and Tamil Nadu, 2008)
- Assisting in the organization of anti-incinerator demonstration in Parma, Italy (2010)
- Creating the anti-incinerator Mesoamerican Network (2015)
- Conducting the fifteenth annual GAIA meeting (2015).

(For a complete list of achievements, see https://www.no-burn.org/history-and-victories/.)

One of the most useful sections of the organization's website is its "Publication and Factsheets" page. That page provides links to several specialized topics, such as cement kilns; waste-to-energy myth; toxic materials; plastic and stopping plastic pollution; landfills; extractive economy; waste and climate; introduction to zero waste; product redesign and producer responsibility; zero waste case studies; medical waste management; organics, compost, and biodigestion; reuse, repair, and recycle; recycling workers and wastepicker's rights; environmental justice; coalitions and community-to-community support; and waste and justice.

Hazel Johnson (1935–2011)

In 1969, Hazel Johnson's husband died of lung cancer at the age of forty-one. Instead of just acknowledging her sorrow and going on with her life, Johnson began to question the cause of death of such a young man. The more she asked, the more she learned about the unhealthy housing development—Altgeld Gardens—in which she lived. She soon found out that Altgeld Gardens was at the center of a "toxic doughnut" with the highest incidence of cancer of any neighborhood in Chicago. The "toxic doughnut" consisted of a ring more than 140 miles in circumference of incinerators, water and sewage treatment plants, steel mills, paint factories, scrapyards, more than fifty landfills, and three abandoned wastewater lagoons. The effluents from these facilities appeared to be responsible for the astonishing number of health problems—asthma and other allergies, excessive nosebleeds and headaches, dizziness and fainting spells, and high rates of cancer—recorded for Altgeld Gardens.

By 1979, Johnson had begun to organize her community to learn more about the health problems they faced and to do something about those problems. The organization she formed, People for Community Recovery (PCR), was eventually incorporated in 1982. Conducting door-to-door surveys of health issues among local residents, PCR volunteers found that more than 90 percent of the 10,000 residents of Altgeld Gardens reported one health problem or more. When Johnson presented these results to city officials, they rejected pending plans for the construction of a new chemical waste incinerator in the area and decided not to expand existing landfills, as they had planned to do.

Johnson and the PCR have realized a number of other successes. They organized a campaign to have asbestos removed from resident apartments and local schools, developed a program on lead poisoning in the community, created a special study group to deal with a sodium oxide spill in the community, monitored the cleanup of PCB contamination in the area, and sponsored a number of regional conferences on environmental justice issues. Hazel Johnson later ceded many leadership responsibilities at PCR to her daughter, Cheryl.

For her many accomplishments, Hazel Johnson was named the Mother of the Environmental Justice Movement at the First National People of Color Environmental Leadership Summit in 1991. In December 2006, Illinois lieutenant governor Pat Quinn awarded Johnson a state Environmental Hero Award in recognition of her commitment to environmental health and protection. Johnson has also accepted a gold medal for the President's Environment and Conservation Challenge Award on behalf of PCR from President George H. W. Bush. In 2011, the Illinois General Assembly designated a section of 130th Street in Chicago as "Hazel Johnson EJ Way." Johnson died in Chicago on January 12, 2011, at the age of seventy-five.

Keep America Beautiful

The early 1950s saw a modest revolution in American society. A postwar economy brought significant economic success to countless numbers (but certainly not all) of American citizens. Among the fallouts of this economic boom was the development of a "throwaway" way of life. Companies made products that were not designed to be fixed and reused, but to be thrown out once they broke down. And "thrown out" was the right term for the fate of much of the nation's solid wastes. It was dumped into the nearest river, the local trash heap, or some other convenient place. The "convenient place" increasingly turned out to be along the highway. With the creation and development of the national Interstate Highway System, first authorized by Congress in 1956, roadside litter became one of the newest and most common forms of "look away and just throw it away."

Out of this scenario grew a reform movement designed to make Americans aware of the nation's litter problem and to change their habits about personal wastes. One manifestation of that movement was the creation in 1953 of Keep America Beautiful (KAB)—a consortium of American businesses, nonprofit organizations, government agencies, and concerned individuals. Some of the original corporate members were Anheuser-Busch, Coca-Cola, PepsiCo, and Philip Morris.

KAB was and always has been primarily a public education program, with television ads, brochures, and other forms of media designed to make Americans aware of the nation's litter problem, put forth ideas for recycling, and in general, inform the general public how it can contribute to reduction in solid wastes.

One of the earliest campaigns, beginning in 1956, was called "Every Litter Bit Hurts." It featured a little girl, Susan Spotless, who asked "Don't grown-ups know? Every litter bit hurts." The program shifted into high gear in 1960 when the Ad Council, the originator of public-service advertising (PSA), joined forces with KAB to develop new and interesting campaigns about solid waste litter. Among the most famous of these campaigns was the "Crying Indian" PSA, in which Iron Eyes Cody, dressed as a Native American, is seen weeping over the damage caused by waste pollution to the natural environment. The ad was part of the "People Start Pollution. People Can Stop It" campaign, which was launched on Earth Day 1971. Also of significance was the Clean Community System, begun in 1975, which focused on community efforts to deal with solid waste pollution in individual cities and towns. At its peak, an estimated 580 communities had become part of the campaign, which claimed an 88 percent reduction in litter in those communities.

Currently, KAB continues its tradition of organizing campaigns over special topics relating to litter pollution. It now makes greater use of social media, such as MySpace, Instagram, Facebook, and AOL. For example, KAB launched a new program featuring KAB Man, a superhero-type character intended to reach the hard-to-reach twentysomethings age group. Some other recent programs are focused on telephone book recycling, cigarette litter prevention, graffiti cleanup, "cans for cash" recycling, and a "wipe out wireless waste" program. A more complete list of KAB programs and campaigns is available on the organization's website at https://www.kab.org/about-us/mission-history and https://www.kab.org/our-programs.

Charles J. Moore (dates unavailable)

Moore has become famous because of a single very special event in his life: the discovery of the Great Pacific Garbage Patch in the Pacific Ocean in 1997. Moore grew up in Long Beach, California, home to his family for three generations. His father was an industrial chemist by occupation and an avid sailor by off-work hobby. He often took his family to sea to destinations ranging from Guadalupe Island, Mexico, to points in Hawaii. Moore attended the University of San Diego (USD), where he majored in chemistry, while also studying mathematics, physics, and Spanish literature.

After leaving USD, he opened his own woodworking and finishing shop, a business he ran for twenty-five years with considerable success. In 1994, he left the business to create a new oceanographic entity, the Algalita Marine Research Foundation, in Long Beach. The mission of the foundation was to carry out research on the coastal waters of Southern California and work for the restoration of that natural environment. He soon became involved in a variety of oceanographic programs, first as a coordinator of the State Water Resources Control Board's Volunteer Water Monitoring Steering Committee and later as a member of the Southern California Water Research Projects' Bight '98 Steering Committee. In the latter position, he provided his own vessel as a platform for researchers conducting an assessment of the Southern California coastline from Point Conception to Ensenada. As part of that project, he also was instrumental in the development of a system for measuring microplastic particles that is still in use worldwide today.

In 1995, Moore built a new research vessel, the *Alguita*, which he first launched from Hobart, Tasmania. On its maiden voyage, the *Alguita* was damaged and had to return to Long Beach for repairs. When that work was completed, he decided to test the vessel by entering the Transpacific Yacht Race, from Los Angeles to Honolulu. After completing the race, he decided to take a "leisurely" route back to California, a trip that took him through a region known as the North Pacific Subtropical Gyre. The North Pacific Subtropical Gyre is a region in the Pacific Ocean ranging from the western coast of the United States to the western coast of Central America, to the Philippine Islands and Japan. Its total size is just less than 8 million square miles. It was within this area that Moore, for the first time, saw a phenomenon now known as the Great Pacific Garbage Patch (GPGP). It was an area, Moore later said, was notable not for "continuous scraps as it is now, but stray pieces here and there." "I could stand on deck for five minutes," he went on, "seeing nothing but the detritus of civilization in the remotest part of the great Pacific Ocean" ("Profile." n.d. http://www.captain-charles-moore.org/about. Accessed on August 2, 2019).

Moore's experience in the GPGP transformed his life. He decided that he had to devote his work not just to the California coast, but to the much larger problem of plastic pollution in the ocean. The Alguita Foundation has now sponsored more than ten trips into the North Pacific Gyre to learn more about the extent and composition of the GPGP. Moore himself has written two scholarly papers on the topic and has become a speaker in worldwide demand on the topic.

National Recycling Coalition

The National Recycling Coalition (NRC) was formed in 1978 by a group of volunteers who were eager to better inform Americans about the need for and value of recycling as a component of waste management programs. At the time, the organization consisted of nineteen staff members and an operating budget of $2 million. It rapidly grew in size and influence, taking a number of steps that pushed recycling to the forefront of Americans' awareness of waste issues. By the early 2000s, it had developed a membership of more than 4,500 individuals and had created subsidiary organizations to deal with specific aspects of recycling. An example is the Rural Recycling Council, designed to assist rural areas with their unique problems of waste recycling. Another important group was the Recycling Advisory Council (RAC), created in 1989. Partially funded by the EPA, the RAC's mission was to "examine the current status of recycling in the United States with the aim of recommending consensus public policies and private initiatives to increase recycling, consistent with the protection of public health and the environment." The council was staffed by high-ranking members of all phases of the recycling industry in the United States. It eventually issued reports on several recycling topics, such as the definition of recycling, solid waste management costs, and policies and initiatives to promote the recycling of paper and plastics.

By 2009, NRC had fallen on hard times. Interest in recycling appeared to have diminished and agencies from which the coalition received its funding had decreased or ended their donations. The period has since been described as "frenetic and dark days where talks of mergers and bankruptcy were rampant" (Hollis, Bob. 2014. "NRC/RONA Partnership." National Recycling Coalition. https://nrcrecycles.org/nrc-rona-partnership/?doing_wp_cron=1564845511.8134200572967529296875. Accessed on August 3, 2019). At first, bankruptcy appeared to be the only option for the organization. Hoping to keep it alive, however, other members suggested a merger with other waste management and recycling groups. A proposal to join with Keep America Beautiful failed, however, and it was not until the mid-2010s that another option—merger with the Recycling Organizations of North America—became a reality. Since that time, NRC has reorganized and become rejuvenated.

As of later 2019, the NRC claimed to have more than 6,000 members representing twenty-three separate recycling organizations. Its mission is to assist those members with the tools and resources they need to advance the role of recycling in today's society, while reaching out to government, corporations, nonprofits, and the general public with information about recycling. It sponsors, cosponsors, or participates in many statewide and regional meetings on recycling principles and practices, as well as conferences on specialized topics, such as electronic waste recycling, paper and plastics recycling, green construction, and curbside recycling. It also cosponsors the annual Resource Recycling Conference along with The Recycling Partnership and Association of Plastic Recyclers. The organization's website also provides links to several important news stories from other sources about recycling.

National Waste & Recycling Association

The National Waste & Recycling Association (NWRA) is a trade association of private-sector waste and recycling organizations, along with manufacturers, service providers, and other entities that do business with those organizations. NWRA was formed in May 1962 by a group of companies involved in solid waste management. For the first six years, the organization was managed by an association management firm in Chicago. In January 1968, it decided to take over its own operation, moving its headquarters to Washington, DC, and formally taking on the name of National Solid Wastes Management Association (NSWMA). At that time, the organization consisted of three paid employees, led by Executive Director Harold Gershowitz.

In 1997, a somewhat involved reorganization took place. A separate entity, the Waste and Equipment Technology Association (WASTEC) was created, and both NSWMA and WASTEC become members of an umbrella group, the Environmental Industry Association (EIA). That arrangement remained in place for fifteen years. Then, in 2012, the three groups merged, largely as a way of erasing a fair level of confusion as to the relationship of the three organizations, resulting in the formation of the National Waste & Recycling Association, a name by which it is known today.

Much of the work of NWRA is carried out through five regional chapters: Midwest Region, Northeast Region, Sunbelt Region, Virginia, and Pennsylvania. Each region sponsors several special events throughout the year, such as state and regional conferences, specialty meetings, and social events. The national association also sponsors general and specialized meetings, such as its annual convention, WasteExpo, and regional and national conferences on topics such as the intersection of safety and automation, public policy leadership, and health-care waste.

The three pillars of NWRA activities are advocacy, safety, and education. Its advocacy arm is designed to promote policies that benefit the solid waste industry as well as society at large. It focuses on topics such as new vehicle technology, recycling, health-care waste, sustainable materials management and landfills, employee safety, employee shortage, and joint employer issues. Among the most important of its safety activities is development and promotion of the ANSI (American National Standards Institute) Z245 standards for all aspects of the solid waste management industry. Those standards cover virtually every aspect of the industry, including mobile refuse collection and transportation equipment, stationary compactors, waste container safety, facility safety, baling equipment, waste container compatibility dimensions, size-reduction equipment, and landfill operations safety requirements. The organization also sponsors several special events promoting safety within the industry, such as the "Sam Safety" program, frequent "Safety Stand Down" campaigns, the "Slow Down to Get Around" program, and regional "Safety Professional Development" programs.

NWRA educational efforts are aimed primarily at members of the association and industry employees. One key component of those efforts are regular webinars on topics of interest to those individuals, including OSHA record keeping and compliance, property insurance, motor vehicle inspections, and roadway safety. The organization also provides opportunities for members to earn certificates in FILA (Future Industry Leadership Alliance) status and landfill operation.

Nuclear Regulatory Commission

The post–World War II years were marked by dramatic debates over the future of nuclear power in the United States. On one hand, controversy raged over continuation of research on nuclear weapons, such as the fission bombs that ended the war and possible fusion versions of such weapons. In another related, but quite different field, politicians, scientists, and other policymakers argued over the development and control of nuclear power as a possible peacetime source of energy with seemingly unlimited potential.

The first step in resolving these debates occurred in 1946 when the U.S. Congress passed the Atomic Energy Act of that year. The purpose of the act was "to promote the utilization of atomic energy for peaceful purposes to the maximum extent consistent with the common defense and security and with the health and safety of the public." The plans outlined by this act were revised and expanded with adoption of the Atomic Energy Act of 1954. One important feature of that act was creation of the Atomic Energy Commission (AEC). That agency was given wide-ranging authority to promote the development of atomic energy for commercial use, along with requirements for close and careful monitoring of the safety of such facilities.

A conflict of interest seemed apparent to many observers early on in this process. Asking the AEC to promote the development of atomic energy while also having to maintain rigorous safety controls over the operation of those plants seemed inconsistent and contradictory. In 1974, therefore, Congress included in the Energy Reorganization Act of 1974 a provision for the establishment of a new agency, the Nuclear Regulatory Commission. This action separated the promotional aspects of nuclear energy and appropriate regulatory functions. The first of these functions was transferred to a new agency, the Energy Research and Development Agency (later part of the Department of Energy), and the latter, to the Nuclear Regulatory Commission.

Today, Nuclear Regulatory Commission's activities fall into three general areas: reactors, materials, and wastes. The commission is responsible for inspecting and licensing all new nuclear reactors, inspecting and relicensing existing plants, conducting ongoing investigations of operating reactors, and conducting research and testing on new methods and materials for use in reactors. It also monitors the wider use of nuclear materials, as in medical applications, research studies, and industrial applications. Finally, it is responsible for monitoring all relevant information about nuclear wastes, such as where such wastes are being produced, how much waste is generated by each facility, and where and how that waste is being stored. On-site visits to waste depositories are designed to find any problems with storage facilities or other conditions that might pose a threat to human health or the environment. The commission is also responsible for overseeing all activities associated with the decommissioning and closure of retired nuclear power plants. Some aspects of nuclear wastes are the responsibility of other federal agencies, such as the U.S. Environmental Protection Agency, Food and Drug Administration, and state governments.

Examples of the types of activities in which Nuclear Regulatory Commission engages for radioactive waste safety are:

- Determining dose limits for radiation workers and members of the public
- Deciding on exposure limits for individual radionuclides
- Monitoring the use and labeling of radioactive materials
- Posting signs in and around radiation areas
- Reporting the theft or loss of radioactive material
- Applying penalties for not complying with Nuclear Regulatory Commission's regulations

Monitoring of uranium mining tailings is of special concern for Nuclear Regulatory Commission inspectors. About 86 percent of the material removed during uranium mining are left behind as tailings. At one time, little or no attention was paid to these tailings; they were simply piled into heaps adjacent to the mine itself. Better information about uranium by-products has changed this point of view. Experts now realize that uranium also contains solid radioactive materials, such as isotopes of radium and thorium, as well as gaseous radioactive isotopes, especially those of radon. Special disposal devices and procedures are needed, therefore, to prevent the released of radioactivity from all these sources at mine sites (where tailings are usually still stored). This can mean that the model for a successful sanitary landfill has to be modified with a cover, for example, that is impermeable to radiation.

Boyan Slat (1994–)

One of the most extensive problems of solid waste management in the world today is the presence and persistence of plastics in the oceans. The most dramatic example of this problem is the Great Pacific Garbage Patch (GPGP), a region of the Pacific Ocean that may be at least as large as twice the size of France. Scholars have been thinking about possible ways of getting rid of the GPGP ever since its discovery in 1997. Most solutions are based on ways of attacking plastic pollution at its source, by reducing the amount of plastic produced around the world, developing mechanisms for degrading plastics before they are deposited into waterways, and educating the general public about the risks of ocean pollution by plastics.

None of these ideas help with the ocean plastic pollution that already exists. It would be helpful if the rate of plastic disposal could be reduced in the future, but what about the vast quantities of plastic currently present in the ocean? Some ideas have been suggested for designing giant vacuum-like devices that could suck plastics out of the oceans or finding and using microorganisms that can degrade plastics (few of which are currently known). In 2011, a Dutch teenage boy, Boyan Slat, came up with a plan for removing in situ plastic wastes from the oceans. He first hit upon the idea while taking diving lessons on the Greek island of Lesbos in 2010. During his classes, he noted that he actually saw more plastic waste in the oceans than he did fish. This struck him as an idea for a high school science project, which he completed the following year.

The project was based on a system of floating plastic tubes capable of collecting wastes and transferring them to a central storage and disposal site. The idea was a great success; it won the local science fair award, and Slat decided to continue his studies at the Delft Technical University (TUDelft) in aerospace engineering. His plans were sidetracked, however, when he was invited to give a TED talk about his invention. Although his Delft audience was relatively small, video of the talk went viral, and Slat soon became something of a celebrity in the field of plastic waste management. He decided to withdraw from TUDelft and focus all his attention on the GPGP cleansing scheme. In 2013, he founded The Ocean Cleanup, an organization of which he is now CEO (chief executive officer). The organization got its start by way of a crowdfunding campaign that produced $2.2 million from 38,000 donors in 160 countries.

Slat was born in Delft, The Netherlands, on July 27, 1994. His father was an artist, and his mother worked in the tourism industry. His parents were divorced when he was still a baby, and he continues to live in Delft as of 2019. He has already begun to amass awards for his ideas for plastic cleanup. In 2014, he received the United Nations' Champion of the Earth award, the youngest recipient ever for that honor. Among his other honors include the Young Entrepreneur Award for 2015 from King Harald of Norway, a Thiel fellowship endowed by PayPal cofounder Peter Thiel, a European of the Year award from *Reader's Digest* magazine in 2017, and a European Entrepreneur of the Year award for 2018 by the television news network *Euronews.*

Solid Waste Association of North America

The Solid Waste Association of North America (SWANA) dates its origins to December 1961, when six men involved in the solid waste management industry met to discuss the formation of a group through which common problems and other topics of interest could be discussed. A year later, the group was formally organized under the name of the Governmental Refuse Collection and Disposal Association (GRCDA). The new organization grew rapidly, forming other chapters in the north central (1963), central, and northern parts of the state (1964). A decade later, the organization had expanded to form its first chapter outside the state, the Pacific Northwest chapter in Washington (1973) and Oregon (1970). The growing environmental movement of the 1970s saw other parts of the country establishing their own chapters of SWANA, including Canadian Prairie, Ontario, New Mexico, and Utah (1976), Florida and Oklahoma (1978), and Illinois and Texas (1979). Today, the organization claims more than 10,000 members in forty-five chapters in the United States, Canada, and the Caribbean. The organization's formal mission statement states that SWANA is an organization of professionals "committed to advancing from solid waste management to resource management through their shared emphasis on education, advocacy and research" (https://swana.org/AboutSWANA.aspx).

SWANA conducts and sponsors a full range of solid waste associated programs and events. Of primary importance, and occupying its own page on the organization's website, is safety. Many of SWANA's activities are aimed at having waste collection removed from the federal government's list of ten most dangerous jobs in the country (it ranks fifth on the list). Among the programs created to achieve this goal are the following:

- Hauler Safety Outreach, designed to make workers more aware of the accident risks they face and methods to avoid harm as a result of those risks;
- "Slow Down to Get Around," aimed at the general public as a way of reminding them of the driving risks associated with following refuse collection vehicles;
- SWANA Chapter Safety Ambassadors, a program in which each chapter designates a specific individual to speak and

teach about safety risks that are inherently a part of the solid waste management industry;

- "5 to Stay Alive," a campaign to make solid waste workers aware of the specific safety risks they face and ways of dealing with those risks;
- "Safe + Sound," a program developed by the U.S. Occupational Health and Safety Administration that includes webinar and other educational programs as well as local events about safety practices in the industry;
- "Safety Summit," a group of presentations offered at the industry's annual national convention.

SWANA offers a large number of special events of interest to those in the field of solid wastes, averaging about three or four events every week. Some examples include chapter meetings and conferences; new member orientations (via webinars); the annual national convention, WASTECON, as well as regional meetings, such as the Quad State Conference; and regular safety meetings and conferences. The organization's own annual national convention is called SWANApalooza and involves educational sessions, major speeches, and a large exhibition of equipment and materials.

SWANA also has an ambitious educational and training program that includes training courses at chapter, regional, and national meetings; online webinars; special safety training courses; and a full course catalog that lists courses on planning and management, landfill and landfill gas, recycling and special wastes, and collection and transfer. The organization also offers live and online courses that lead to certification in a variety of fields, such as managing leachate recirculation and bioreactor landfills, household hazardous waste and CESQG (conditionally exempt small quantity generator) collection operations, manager of landfill operations, managing MSW collection systems, managing composting programs, managing construction and demolition materials, managing integrated solid waste management systems, managing recycling systems, managing transfer station systems, and zero waste principles and practices.

Beverly Wright (1947–)

A member of the Michigan Coalition, the National Advisory Committee for the First National People of Color Environmental Leadership Summit, the Planning and Protocol Committees for the National Institute for Environmental Health Sciences' Health and Research Needs to Ensure Environmental Justice National Symposium, and the Environmental Protection Agency's National Environmental Justice Advisory Council and founder and director of the Deep South Center for Environmental Justice: these are some of the major activities in which Beverly Hendrix Wright has been engaged over the past three decades. She has also been active in other professional activities sponsored by the National Council for Negro Women, the Association of Black Psychologists, the Association of Social and Behavioral Scientists, and the Society for the Study of Social Problems.

Wright was born in New Orleans on October 1, 1947. She earned her BA degree in sociology from Grambling College in 1969 and her MA (1971) and PhD (1977), both in sociology, from the State University of New York at Buffalo (SUNY-B). She served as instructor in the departments of sociology at Millard Fillmore College (1970–1971) and SUNY-B (1970–1974). Wright was appointed to the faculty at the University of New Orleans in 1974 and then was promoted to assistant professor in 1977 and to associate professor in 1987. During the period from 1989 to 1993, she served as associate professor of sociology at Wake Forest University. In 1992, she founded the Deep South Center for Environmental Justice (DSCEJ) at Xavier University of Louisiana in New Orleans. She was named director of the center, a post she continues to hold.

In 2005, DSCEJ made plans to move its operations from Xavier to Dillard University, across town in New Orleans. That move was scheduled to take place on September 1 but was disrupted when Hurricane Katrina struck the city three days earlier. In spite of the chaos and destruction caused by the hurricane, the move eventually took place, and the center is now at its new home at Dillard. One of the major projects on which it has focused its attention in the past few years has been on assisting in the city's recovery from one of the nation's greatest natural disasters.

In addition to her work at DSCEJ, Wright has served on a number of community and professional committees and commissions, including the New Orleans Mayor's Environmental Advisory Committee, the Mayor's Office of Environmental Affairs Brownfields Consortium, the New Orleans' Mayor's Committee on Solid Waste, the Army Corps of Engineers' Environmental Advisory Board, and the U.S. Commission of Civil Rights for the State of Louisiana. She is the author or coauthor of more than two dozen papers, reviews, and book chapters. Among her recent books are *Deadly Waiting Game beyond Katrina: How Government Actions Endanger the Health and Welfare of African Americans* (with Robert Bullard, New York University Press, 2011) and *Race Place and Environmental Justice in the Aftermath of Katrina: Struggles to Reclaim, Rebuild and Revitalize New Orleans and the Gulf Coast* (with Robert Bullard Boulder, Colorado: Westview Press, 2009).

Wright has received a number of awards and honors, including a Special Gulf Coast Award for outstanding leadership in the aftermath of Hurricane Katrina from the Robert Wood Johnson Community Health Leadership Program in 2006; Distinguished Alumni Award in 2012 from the Sociology Department, University at Buffalo, State University of New York; SAGE Activist Scholar Award in 2011 from the Urban Affairs Association; Conrad Arensberg Award in 2010 from the Society for the Anthropology of Work; Beta Kappa Chi Humanitarian Assistance Award in 2010 from the National Institutes of Science; Heinz Award in 2009 from the Heinz Family Foundation; Freedom Sisters Award in 2009 from the Ford Motor Company; and the Environmental Justice Achievement Award in 2008 from the U.S. Environmental Protection Agency. She continues to serve as executive director of the Deep South Center for Environmental Justice and professor of sociology at Dillard University.

roid International

mical Corporation

Introduction

This chapter provides data on the materials in the waste stream and how they were dealt with (recycled, composted, incinerated, or landfilled). It also provides a list of Superfund sites and an analysis of hazardous waste generation. This data is followed by excerpts from primary documents, including laws, court decisions, and other materials.

Data

A leaking barrel of industrial waste. If not disposed of responsibly, toxic and other hazardous industrial waste may leach into the soil and contaminate groundwater. (Corel)

Table 5.1 Materials Generated in the Municipal Waste Stream, 1960–2015 (thousands of tons)

The U.S. Environmental Protection Agency collects and provides statistics on many aspects of waste management in the United States annually. This table provides data on the types and amounts of wastes generated in the United States from 1960 to 2014.

Materials	1960	1970	1980	1990	2000	2005	2010	2014	2015
Paper and Paperboard	29,990	44,310	55,160	72,730	87,740	84,840	71,310	68,610	68,050
Glass	6,720	12,740	15,130	13,100	12,770	12,540	11,520	11,480	11,470
Metals									
Ferrous	10,300	12,360	12,620	12,640	14,150	15,210	16,920	17,880	18,170
Aluminum	340	800	1,730	2,810	3,190	3,330	3,510	3,530	3,610
Other Nonferrous	180	670	1,160	1,100	1,600	1,860	2,020	2,230	2,220
Total Metals	10,820	13,830	15,510	16,550	18,940	20,400	22,450	23,640	24,000
Plastics	390	2,900	6,830	17,130	25,550	29,380	31,400	33,390	34,500
Rubber and Leather	1,840	2,970	4,200	5,790	6,670	7,290	7,750	8,210	8,480
Textiles	1,760	2,040	2,530	5,810	9,480	11,510	13,220	15,240	16,030
Wood	3,030	3,720	7,010	12,210	13,570	14,790	15,710	16,120	16,300
Other	70	770	2,520	3,190	4,000	4,290	4,710	5,120	5,160
Total Materials in Products	54,620	83,280	108,890	146,510	178,720	185,040	178,070	181,810	183,990
Other Wastes									
Food	12,200	12,800	13,000	23,860	30,700	32,930	35,740	38,670	39,730
Yard Trimmings	20,000	23,200	27,500	35,000	30,530	32,070	33,400	34,500	34,720
Miscellaneous Inorganic Wastes	1,300	1,780	2,250	2,900	3,500	3,690	3,840	3,970	3,990
Total Other Wastes	33,500	37,780	42,750	61,760	64,730	68,690	72,980	78,440	77,140
Total MSW Generated	88,120	121,060	151,640	208,270	243,450	253,730	251,050	258,950	262,430

Source: "Materials Generated in the Municipal Waste Stream, 1960 to 2015." 2018. "Advancing Sustainable Materials Management: 2015 Tables and Figures." Table 1. Washington, DC: Environmental Protection Agency. https://www.epa.gov/facts-and-figures-about-materials-waste-and-recycling/advancing-sustainable-materials-management. Accessed on June 15, 2019.

Table 5.2 **Materials Generated in the Municipal Waste Stream, 1960–2015**

This table follows up on Table 5.1, providing the percentage of wastes coming from about a dozen major sources.

Materials	1960	1970	1980	1990	2000	2005	2010	2014	2015
Paper and Paperboard	34.0%	36.6%	36.4%	34.9%	36.0%	33.4%	28.4%	26.5%	25.9%
Glass	7.6%	10.5%	10.0%	6.3%	5.2%	4.9%	4.6%	4.4%	4.4%
Metals									
Ferrous	11.7%	10.2%	8.3%	6.1%	5.8%	6.0%	6.7%	6.9%	6.9%
Aluminum	0.4%	0.7%	1.1%	1.3%	1.3%	1.3%	1.4%	1.4%	1.4%
Other Nonferrous	0.2%	0.6%	0.8%	0.5%	0.7%	0.7%	0.8%	0.8%	0.8%
Total Metals	12.3%	11.4%	10.2%	7.9%	7.8%	8.0%	8.9%	9.1%	9.1%
Plastics	0.4%	2.4%	4.5%	8.2%	10.5%	11.6%	12.5%	12.9%	13.1%
Rubber and Leather	2.1%	2.5%	2.8%	2.8%	2.7%	2.9%	3.1%	3.2%	3.2%
Textiles	2.0%	1.7%	1.7%	2.8%	3.9%	4.5%	5.3%	5.9%	6.1%
Wood	3.4%	3.1%	4.6%	5.9%	5.6%	5.8%	6.3%	6.2%	6.2%
Other	0.1%	0.6%	1.7%	1.5%	1.6%	1.7%	1.9%	2.0%	2.1%
Total Materials in Products	62.0%	68.8%	71.8%	70.3%	73.4%	72.9%	70.9%	70.2%	70.1%
Other Wastes									
Food	13.8%	10.6%	8.6%	11.5%	12.6%	13.0%	14.2%	14.9%	15.1%
Yard Trimmings	22.7%	19.2%	18.1%	16.8%	12.5%	12.6%	13.3%	13.3%	13.3%
Miscellaneous Inorganic Wastes	1.5%	1.5%	1.5%	1.4%	1.4%	1.5%	1.5%	1.5%	1.5%
Total Other Wastes	38.0%	31.2%	28.2%	29.7%	26.6%	27.1%	29.1%	29.8%	29.9%
Total MSW Generated	100%	100%	100%	100%	100%	100%	100%	100%	100%

Source: "Materials Generated in the Municipal Waste Stream, 1960 to 2015." 2018. "Advancing Sustainable Materials Management: 2015 Tables and Figures." Table 1. Washington, DC: Environmental Protection Agency. https://www.epa.gov/facts-and-figures-about-materials-waste-and-recycling/advancing-sustainable-materials-management. Accessed on June 15, 2019.

Table 5.3 Municipal Solid Wastes Recycled and Composted, 1960–2015

This table covers data similar to that of Table 5.1 and 5.2, showing the amounts of municipal solid waste either recycled or composted.

Materials	1960	1970	1980	1990	2000	2005	2010	2014	2015
Paper and Paperboard	5,080	6,770	11,740	20,230	37,560	41,960	44,570	44,400	45,320
Glass	100	160	750	2,630	2,880	2,590	3,130	2,990	3,030
Metals									
Ferrous	50	150	370	2,230	4,680	5,020	5,800	5,970	6,060
Aluminum	neg.	10	310	1,010	860	690	680	710	670
Other Nonferrous	neg.	320	540	730	1,060	1,280	1,440	1,550	1,500
Total Metals	50	480	1,220	3,970	6,600	6,990	7,920	8,230	8,230
Plastics	neg.	neg.	20	370	1,480	1,780	2,500	3,190	3,140
Rubber and Leather	330	250	130	370	820	1,050	1,440	1,440	1,510
Textiles	50	60	160	660	1,320	1,830	2,050	2,260	2,450
Wood	neg.	neg.	neg.	130	1,370	1,830	2,280	2,570	2,660
Other	neg.	300	500	680	980	1,210	1,370	1,470	1,430
Total Materials in Products	5,610	8,020	14,520	29,040	53,010	59,240	65,260	66,550	67,770
Other Wastes									
Food	neg.	neg.	neg.	neg.	680	690	970	1,940	2,100
Yard Trimmings	neg.	neg.	neg.	4,200	15,770	19,860	19,200	21,080	21,290
Miscellaneous Inorganic Wastes	neg.	neg.	neg.	neg.	neg.	neg.	neg.	neg.	neg.
Total Other Wastes	neg.	neg.	neg.	4,200	16,450	20,550	20,170	23,020	23,390
Total MSW Recycled and Composted	5,610	8,020	14,520	33,240	69,460	79,790	85,430	89,570	91,160

neg. = negligible.

Source: "Materials Recycled and Composted in the Municipal Waste Stream, 1960 to 2015." 2018. "Advancing Sustainable Materials Management: 2015 Tables and Figures." Table 2. Washington, DC: Environmental Protection Agency. https://www.epa.gov/sites/production/files/2018 07/documents/smm_2015_tables_and_figures_07252018_fnl_508_0.pdf. Accessed on June 15, 2019.

Table 5.4 Materials Combusted with Energy Recovery, 1970–2015

This table continues coverage similar to that for Tables 5.1, 5.2, and 5.3, showing the amount of municipal solid waste incinerated for the purpose of producing energy.

Materials	1970	1980	1990	2000	2005	2010	2014	2015
Paper and Paperboard	150	860	8,930	9,730	7,800	4,740	4,740	4,450
Glass	60	300	1,810	1,790	1,660	1,360	1,450	1,470
Metals								
Ferrous	60	250	1,690	1,610	1,640	1,810	2,030	2,140
Aluminum	0	30	300	390	410	440	470	500
Other Nonferrous	0	20	60	50	50	60	50	60
Total Metals	60	300	2,050	2,050	2,100	2,310	2,550	2,700
Plastics	0	140	2,980	4,120	4,330	4,530	5,010	5,350
Rubber and Leather	10	70	830	1,970	2,110	1,910	2,620	2,490
Textiles	10	50	880	1,880	2,110	2,270	3,020	3,050
Wood	10	150	2,080	2,290	2,270	2,310	2,540	2,580
Other	0	30	410	540	510	540	670	690
Total Materials in Products	300	1,900	19,970	24,370	22,890	19,970	22,600	22,780
Other Wastes								
Food	50	260	4,060	5,820	5,870	6,150	7,200	7,380
Yard Trimmings	90	550	5,240	2,860	2,220	2,510	2,630	2,630
Miscellaneous Inorganic Wastes	10	50	490	680	670	680	780	780
Total Other Wastes	150	860	9,790	9,360	8,760	9,340	10,610	10,790
Total MSW Combusted	450	2,760	29,760	33,730	31,650	29,310	33,210	33,570

Source: "Materials Generated in the Municipal Waste Stream, 1960 to 2015." 2018. "Advancing Sustainable Materials Management: 2015 Tables and Figures." Table 3. Washington, DC: Environmental Protection Agency. https://www.epa.gov/facts-and-figures-about-materials-waste-and-recycling/advancing-sustainable-materials-management. Accessed on June 15, 2019.

Table 5.5 Materials Landfilled in the Municipal Waste Stream, 1960–2015

This table provides additional information about the amounts of solid wastes deposited in landfills from 1960 to 2015 (thousands of tons).

Materials	1960	1970	1980	1990	2000	2005	2010	2014	2015
Paper and Paperboard	24,910	37,390	42,560	43,570	40,450	35,080	22,000	19,470	18,280
Glass	6,620	12,520	14,080	8,660	8,100	8,290	7,030	7,040	6,970
Metals									
Ferrous	10,250	12,150	12,000	8,720	7,860	8,550	9,310	9,880	9,970
Aluminum	340	790	1,390	1,500	1,940	2,230	2,390	2,350	2,440
Other Nonferrous	180	350	600	310	490	530	520	630	660
Total Metals	10,770	13,290	13,990	10,530	10,290	11,310	12,220	12,860	13,070
Plastics	390	2,900	6,670	13,780	19,950	23,270	24,370	25,190	26,010
Rubber and Leather	1,510	2,710	4,000	4,590	3,880	4,130	4,400	4,150	4,480
Textiles	1,710	1,970	2,320	4,270	6,280	7,570	8,900	9,960	10,530
Wood	3,030	3,710	6,860	10,000	9,910	10,690	11,120	11,010	11,060
Other	70	470	1,990	2,100	2,480	2,570	2,800	2,980	3,040
Total Materials in Products	49,010	74,960	92,470	97,500	101,340	102,910	92,840	92,660	93,440
Other Wastes									
Food	12,200	12,750	12,740	19,800	24,200	26,370	28,620	29,530	30,250
Yard Trimmings	20,000	23,110	26,950	25,560	11,900	9,990	11,690	10,790	10,800
Miscellaneous Inorganic Wastes	1,300	1,770	2,200	2,410	2,820	3,020	3,160	3,190	3,210
Total Other Wastes	33,500	37,630	41,890	47,770	38,920	39,380	43,470	43,510	44,260
Total MSW Landfilled	82,510	112,590	134,360	145,270	140,260	142,290	136,310	136,170	137,700

Source: "Materials Generated in the Municipal Waste Stream, 1960 to 2015." 2018. "Advancing Sustainable Materials Management: 2015 Tables and Figures." Table 2. Washington, DC: Environmental Protection Agency. https://www.epa.gov/facts-and-figures-about-materials-waste-and-recycling/advancing-sustainable-materials-management. Accessed on June 15, 2019.

Table 5.6 List of Superfund National Priority List Sites, 1983–2019

The status of waste disposal sites on the Environmental Protection Agency's National Priority List is constantly changing. That list includes sites that are eligible for cleanup, where cleanup has been completed, and where partial cleanup has occurred. This table provides data on those sites from 1983 to 2019.

Action	2019	2018	2017	2016	2015	2014	2013	2012	2011	2010	2009	2008	2007	2006
Proposed Sites	2	19	4	16	13	16	9	18	35	8	23	17	17	10
NPL Sites	7	15	7	15	8	21	9	24	25	20	20	18	12	10
Deleted Sites	1	18	2	2	6	14	7	11	7	7	8	9	7	7
Sites with Construction Completions	1	10	7	11	13	8	14	22	22	18	20	30	24	40
Sites with Partial Deletions	4	2	3	0	1	3	1	0	2	3	3	3	2	1
Partial Deletion Actions	4	4	4	1	2	4	2	2	3	5	3	3	3	3

Action	2005	2004	2003	2002	2001	2000	1999	1998	1997	1996	1995	1994	1993	1992
Proposed Sites	12	26	14	9	45	40	37	34	20	27	9	36	52	30
NPL Sites	18	11	20	19	29	39	43	17	18	13	31	43	33	0
Deleted Sites	18	16	9	17	30	19	23	20	32	34	26	13	12	2
Sites with Construction Completions	40	40	40	42	47	87	85	87	88	64	68	61	68	88
Sites with Partial Deletions	5	3	6	5	3	5	3	7	6	0	0	0	0	0
Partial Deletion Actions	5	7	7	7	4	5	3	7	6	0	0	0	0	0

Action	1991	1990	1989	1988	1987	1986	1985	1984	1983
Proposed Sites	23	25	64	246	71	45	317	0	552
NPL Sites	7	300	101	0	99	170	3	132	406
Deleted Sites	9	1	10	5	0	8	0	0	5
Sites with Construction Completions	12	8	10	12	3	8	3	0	5
Sites with Partial Deletions	0	0	0	0	0	0	0	0	0
Partial Deletion Actions	0	0	0	0	0	0	0	0	0

Source: "Number of NPL Site Actions and Milestones by Fiscal Year." 2019. Environmental Protection Agency. https://www.epa.gov/superfund/number-npl-site-actions-and-milestones-fiscal-year. Accessed on June 22, 2019.

Table 5.7 Trend Analysis for Hazardous Waste Generation, 2001–2017, National and by Selected States

Data on hazardous waste generation and disposal are difficult to come by because the federal government relies on reports from individual states, who are responsible for monitoring these sites. These data are among the most easily accessible statistics on the issue.

Location	2001	2003	2005	2007	2009	2011	2013	2015	2017
National (#)[1]	19,019	17,692	16,211	16,386	16,215	16,727	25,221	26,801	25,583
National (wgt)[2]	40,551	30,176	36,860	33,313	35,328	34,835	33,767	33,647	35,101
California (#)	2,544	2,514	2,236	2,310	2,578	1,511	3,768	3,394	3,945
California (wgt)	807,297	445,317	747,233	534,752	696,067	1,203,326	354,288	309,109	347,196
New York (#)	1,993	1,339	1,036	1,181	1,190	1,471	3,094	3,636	2,853
New York (wgt)	3,534,275	1,130,623	1,124,198	1,267,648	1,032,626	186,483	234,393	281,259	145,620
Ohio (#)	1,071	1,040	1,040	980	896	941	1,221	1,330	1,255
Ohio (wgt)	1,889,067	1,800,170	2,198,821	1,612,560	1,300,804	1,627,192	1,531,251	1,711,527	1,594,454
Maryland (#)	14	132	122	117	112	134	466	342	344
Maryland (wgt)	17,577	55,379	39,715	43,606	33,684	44,250	47,987	125,807	41,488
North Dakota (#)	15	18	13	13	14	19	26	35	30
North Dakota (wgt)	574,614	633,735	549,686	538,611	530,504	455,868	375,752	265,051	421,766
South Dakota (#)	16	21	14	19	25	42	40	43	41
South Dakota (wgt)	950	1,254	992	750	1,214	1,347	1,475	1,917	1,966

[1] Number of generators.

[2] Weight generated, in thousands of tons, rounded.

Source: "Trends Analysis Results for National, Generation." 2018. Environmental Protection Agency. https://rcrapublic.epa.gov/rcrainfoweb/action/modules/br/trends. Accessed on June 22, 2019.

Documents

Waste management has long been one of the most contentious legal issues in the United States as well as other parts of the world. Untold number of laws have been passed, administrative rulings issued, and court cases heard on a seemingly endless range of very specific questions: the definition of wastes; their transportation; disposal and storage; different effects on various populations, human health, and the environment; and other issues. The selections included here can do no more than give a taste of this wide variety of legal issues addressed by those laws and court cases.

Resource Conservation and Recovery Act (1976)

The Resource Conservation and Recovery Act of 1976 is a very long and complex act that has been amended several times. It is the principal federal law governing the disposal of solid and hazardous wastes. A key, defining section of the act is section 6902 of the US Code, which lays out the objectives of the act and the national policy on the topic. That section reads as follows:

(a) Objectives

The objectives of this chapter are to promote the protection of health and the environment and to conserve valuable material and energy resources by—

(1) providing technical and financial assistance to State and local governments and interstate agencies for the development of solid waste management plans (including resource recovery and resource conservation systems) which will promote improved solid waste management techniques (including more effective organizational arrangements), new and improved methods of collection, separation, and recovery of solid waste, and the environmentally safe disposal of nonrecoverable residues;

(2) providing training grants in occupations involving the design, operation, and maintenance of solid waste disposal systems;

(3) prohibiting future open dumping on the land and requiring the conversion of existing open dumps to facilities which do not pose a danger to the environment or to health;

(4) assuring that hazardous waste management practices are conducted in a manner which protects human health and the environment;

(5) requiring that hazardous waste be properly managed in the first instance thereby reducing the need for corrective action at a future date;

(6) minimizing the generation of hazardous waste and the land disposal of hazardous waste by encouraging process substitution, materials recovery, properly conducted recycling and reuse, and treatment;

(7) establishing a viable Federal-State partnership to carry out the purposes of this chapter and insuring that the Administrator will, in carrying out the provisions of subchapter III of this chapter, give a high priority to assisting and cooperating with States in obtaining full authorization of State programs under subchapter III of this chapter;

(8) providing for the promulgation of guidelines for solid waste collection, transport, separation, recovery, and disposal practices and systems;

(9) promoting a national research and development program for improved solid waste management and resource conservation techniques, more effective organizational arrangements, and new and improved methods of collection, separation, and recovery, and recycling of solid wastes and environmentally safe disposal of nonrecoverable residues;

(10) promoting the demonstration, construction, and application of solid waste management, resource recovery,

and resource conservation systems which preserve and enhance the quality of air, water, and land resources; and

(11) establishing a cooperative effort among the Federal, State, and local governments and private enterprise in order to recover valuable materials and energy from solid waste.

Source: Resource Conservation and Recovery Act of 1976 (94–580; 90 Stat. 2795). Title 42: The Public Health and Welfare. 2019. U.S. Code. Section 6902, p. 5993.

Margaret Bean et al. v. Southwestern Waste Management Corp. (1979)

Cases involving environmental inequities have sometimes been difficult to pursue, partly because of the U.S. Supreme Court's decision in the case of Washington v. Davis *(426 U.S. 229). In that case, the court ruled that plaintiffs had to provide "discriminatory intent"—that is, that an individual, agency, or corporation specifically* intended *to cause harm to an individual or a group of individuals. The effect of this ruling can be seen in the first suit brought on the basis of disproportionate exposure to environmental hazards,* Bean v. Southwestern. *In this case, plaintiffs attempted to prevent the construction of a solid waste disposal facility in Houston on the grounds that it had a disproportionate impact on the minority community in which it would have been located. The district court refused to grant a temporary injunction but allowed plaintiffs to proceed to the discovery stage of their suit. Eventually, however, a different judge dismissed the suit completely. (Triple asterisks, ***, indicate omission of references.)*

The Preliminary Injunction

There are four prerequisites to the granting of a preliminary injunction. The plaintiffs must establish: (1) a substantial likelihood of success on the merits, (2) a substantial threat of irreparable injury, "(3) that the threatened injury to the plaintiff[s] outweighs the threatened harm the injunction may do to defendant[s], and (4) that granting the preliminary injunction will not disserve the public interest."***

[The court then goes on to say that plaintiffs have satisfied the second of those conditions, but not the first.]

The problem is that the plaintiffs have not established a substantial likelihood of success on the merits. The burden on them is to prove discriminatory purpose.*** That is, the plaintiffs must show not just that the decision to grant the permit is objectionable or even wrong, but that it is attributable to an intent to discriminate on the basis of race. Statistical proof can rise to the level that it, alone, proves discriminatory intent, as in***, or, this Court would conclude, even in situations less extreme than in those two cases, but the data shown here does not rise to that level. Similarly, statistical proof can be sufficiently supplemented by the types of proof outlined in Arlington Heights, supra, to establish purposeful discrimination, but the supplemental proof offered here is not sufficient to do that.

. . .

If this Court were TDH *[Texas Department of Health]*, it might very well have denied this permit. It simply does not make sense to put a solid waste site so close to a high school, particularly one with no air conditioning. Nor does it make sense to put the land site so close to a residential neighborhood. But I am not TDH and for all I know, TDH may regularly approve of solid waste sites located near schools and residential areas, as illogical as that may seem.

It is not my responsibility to decide whether to grant this site a permit. It is my responsibility to decide whether to grant the plaintiffs a preliminary injunction. From the evidence before me, I can say that the plaintiffs have established that the decision to grant the permit was both unfortunate and insensitive. I cannot say that the plaintiffs have established a substantial likelihood of proving that the decision to grant the permit was motivated by purposeful racial discrimination in violation of 42 U.S.C. § 1983. This Court is obligated, as all Courts are, to follow the precedent of the United States Supreme Court and the evidence adduced thus far does not meet the magnitude required by *Arlington Heights, supra.*

Conclusion

At this juncture, the decision of TDH seems to have been insensitive and illogical. Sitting as the hearing examiner for TDH, based upon the evidence adduced, this Court would have denied the permit. But this Court has a different role to play, and that is to determine whether the plaintiffs have established a substantial likelihood of proving that TDH's decision to issue the permit was motivated by purposeful discrimination in violation of 42 U.S.C. § 1983 as construed by superior courts. That being so, it is hereby ORDERED, ADJUDGED, and DECREED that the plaintiffs' Motion for a Preliminary Injunction be, and the same is, DENIED. For the reasons stated above, the defendants' Motions to Dismiss are also DENIED.

Source: *Bean v. Southwestern Waste Management Corp.*, 482 F. Supp. 673 (S.D. Tex. 1979).

Comprehensive Environmental Response, Compensation, and Liability Act (1980)

The 1970s were a decade in the United States of growing awareness of a host of environmental problems that had been developing as the result of rapid technological and industrial development. Several important federal agencies were created, and other state and federal actions taken, to deal with these problems. For example, President Richard M. Nixon created the Environmental Protection Agency (EPA) on December 2, 1970. Waste products released to the environment were a major focus of much of that legislation and administrative action. One of the first pieces of waste legislation to be adopted was the Comprehensive Environmental Response, Compensation, and Liability Act of 1980 (CERCLA). The act created a program known as Superfund to pay for the rehabilitation of sites that had been used for the creation, storage, transportation, and other activity related to hazardous wastes. That act was later amended and updated to keep the program in operation to the present day. As of early 2019, there were still about 1,300 sites listed by the EPA as Superfund projects. The selection below is an excerpt of the very long CERCLA act passed in 1980.

National Contingency Plan

SEC. 105. Within one hundred and eighty days after the enactment of this Act, the President shall, after notice and opportunity for public comments, revise and republish the national contingency plan for the removal of oil and hazardous substances, originally prepared and published pursuant to section 311 of the Federal Water Pollution Control Act, to reflect and effectuate the responsibilities and powers created by this Act, in addition to those matters specified in section 311(c)(2). Such revision shall include a section of the plan to be known as the national hazardous substance response plan which shall establish procedures and standards for responding to releases of hazardous substances, pollutants, and contaminants, which shall include at a minimum:

(1) methods for discovering and investigating facilities at which hazardous substances have been disposed of or otherwise come to be located;

(2) methods for evaluating, including analyses of relative cost, and remedying any releases or threats of releases from facilities which pose substantial danger to the public health or the environment;

(3) methods and criteria for determining the appropriate extent of removal, remedy, and other measures authorized by this Act;

(4) appropriate roles and responsibilities for the Federal, State, and local governments and for interstate and nongovernmental entities in effectuating the plan;

(5) provision for identification, procurement, maintenance, and storage of response equipment and supplies;

(6) a method for and assignment of responsibility for reporting the existence of such facilities which may be located on federally owned or controlled properties and any releases of hazardous substances from such facilities;

(7) means of assuring that remedial action measures are cost effective over the period of potential exposure to the hazardous substances or contaminated materials;

(8)(A) criteria for determining priorities among releases or threatened releases throughout the United States for the purpose of taking remedial action and, to the extent practicable taking into account the potential urgency of such action, for the purpose of taking removal action. Criteria and priorities under this paragraph shall be based upon relative risk or danger to public health or welfare or the environment, in the judgment of the President, taking into account to the extent possible the population at risk, the hazard potential of the hazardous substances at such facilities, the potential for contamination of drinking water supplies, the potential for direct human contact, the potential for destruction of sensitive ecosystems, State preparedness to assume State costs and responsibilities, and other appropriate factors;

(B) based upon the criteria set forth in subparagraph (A) of this paragraph, the President shall list as part of the plan national priorities among the known releases or threatened releases throughout the United States and shall revise the list no less often than annually. Within one year after the date of enactment of this Act, and annually thereafter, each State shall establish and submit for consideration by the President priorities for remedial action among known releases and potential releases in that State based upon the criteria set forth in subparagraph (A) of this paragraph. In assembling or revising the national list, the President shall consider any priorities established by the States. To the extent practicable, at least four hundred of the highest priority facilities shall be designated individually and shall be referred to as the "top priority among known response targets", and, to the extent practicable, shall include among the one hundred highest priority facilities at least one such facility from each State which shall be the facility designated by the State as presenting the greatest danger to public health or welfare or the environment among the known facilities in such State. Other priority facilities or incidents may be listed singly or grouped for response priority purposes; and

(9) specified roles for private organizations and entities in preparation for response and in responding to releases of hazardous substances, including identification of appropriate qualifications and capacity therefor. The plan shall specify procedures, techniques, materials, equipment, and methods to be employed in identifying, removing, or remedying releases of hazardous substances comparable to those required under section 311(c)(2) (F) and (G) and (j)(l) of the Federal Water Pollution Control Act. Following publication of the revised national contingency plan, the response to and actions to minimize damage from hazardous substances releases shall, to the greatest extent possible, be in accordance with the provisions of the plan. The President may, from time to time, revise and republish the national contingency plan.

Source: Comprehensive Environmental Response, Compensation, and Liability Act of 1980. Public Law 96–510. 42 U.S.C. §9601.

Basel Convention on the Control of Transboundary Movements of Hazardous Wastes and Their Disposal (1989)

One of the most serious waste management problems worldwide until just a few decades ago was the shipment of hazardous wastes from one country to another. The transfer most often took place when a developed country, such as the United States, decided to transport wastes that were harmful to the environment or human health to a developing nation, such as Nigeria. In theory, that process was favorable to both parties: the developed nation got rid of wastes that posed a threat to the country, while the developing nation earned substantial international currency to promote its own development. The problem with this arrangement was, of course, the potential harm posed by these wastes to the receiving country, harm that was by no means inconsiderable. On March 22, 1989, at a meeting of the United Nations, the Basel Convention on the Control of Transboundary Movements of Hazardous Wastes and Their Disposal (usually known just as the Basel Convention) was adopted. The purpose of that treaty was to bring under control the transfer of hazardous wastes between nations. As of 2019, there were 187 parties to the treaty, which went into effect on May 5, 1992. A core section of the treaty is as follows:

Article 4
General Obligations

1. (a) Parties exercising their right to prohibit the import of hazardous wastes or other wastes for disposal shall inform the other Parties of their decision pursuant to Article 13.

   (b) Parties shall prohibit or shall not permit the export of hazardous wastes and other wastes to the Parties which have prohibited the import of such wastes, when notified pursuant to subparagraph (a) above.

(c) Parties shall prohibit or shall not permit the export of hazardous wastes and other wastes if the State of import does not consent in writing to the specific import, in the case where that State of import has not prohibited the import of such wastes.

2. Each Party shall take the appropriate measures to:

   (a) Ensure that the generation of hazardous wastes and other wastes within it is reduced to a minimum, taking into account social, technological and economic aspects;

   (b) Ensure the availability of adequate disposal facilities, for the environmentally sound management of hazardous wastes and other wastes, that shall be located, to the extent possible, within it, whatever the place of their disposal;

   (c) Ensure that persons involved in the management of hazardous wastes or other wastes within it take such steps as are necessary to prevent pollution due to hazardous wastes and other wastes arising from such management and, if such pollution occurs, to minimize the consequences thereof for human health and the environment;

   (d) Ensure that the transboundary movement of hazardous wastes and other wastes is reduced to the minimum consistent with the environmentally sound and efficient management of such wastes, and is conducted in a manner which will protect human health and the environment against the adverse effects which may result from such movement;

   (e) Not allow the export of hazardous wastes or other wastes to a State or group of States belonging to an economic and/or political integration organization that are Parties, particularly developing countries, which have prohibited by their legislation all imports, or if it has reason to believe that the wastes in question will not be managed in an environmentally sound manner, according to criteria to be decided on by the Parties at their first meeting;

(f) Require that information about a proposed transboundary movement of hazardous wastes and other wastes be provided to the States concerned, according to Annex V A, to state clearly the effects of the proposed movement on human health and the environment;

(g) Prevent the import of hazardous wastes and other wastes if it has reason to believe that the wastes in question will not be managed in an environmentally sound manner;

(h) Co-operate in activities with other Parties and interested organizations, directly and through the Secretariat, including the dissemination of information on the transboundary movement of hazardous wastes and other wastes, in order to improve the environmentally sound management of such wastes and to achieve the prevention of illegal traffic.

3. The Parties consider that illegal traffic in hazardous wastes or other wastes is criminal.

Source: "Basel Convention on the Control of Transboundary Movements of Hazardous Wastes and Their Disposal." United Nations *Treaty Series*, 1673: 57.

Principles of Environmental Justice (1991)

On October 24–27, 1991, a conference on environmental justice, the National People of Color Environmental Leadership Summit, was held in Washington, DC. At the conclusion of that meeting, participants in the conference adopted a set of principles for their movement, a statement that is still widely regarded as providing a set of guiding principles for the environmental justice movement.

Preamble

WE, THE PEOPLE OF COLOR, gathered together at this multinational People of Color Environmental Leadership Summit, to begin to build a national and international movement of all peoples of color to fight the destruction and taking of our lands and communities, do hereby reestablish our spiritual interdependence to the sacredness of our Mother Earth; to respect and celebrate each of our cultures, languages and beliefs about the natural world and our roles in healing ourselves; to ensure environmental justice; to promote economic alternatives which would contribute to the development of environmentally safe livelihoods; and, to secure our political, economic and cultural liberation that has been denied for over 500 years of colonization and oppression, resulting in the poisoning of our communities and land and the genocide of our peoples, do affirm and adopt these Principles of Environmental Justice:

1) **Environmental Justice** affirms the sacredness of Mother Earth, ecological unity and the interdependence of all species, and the right to be free from ecological destruction.
2) **Environmental Justice** demands that public policy be based on mutual respect and justice for all peoples, free from any form of discrimination or bias.
3) **Environmental Justice** mandates the right to ethical, balanced and responsible uses of land and renewable resources in the interest of a sustainable planet for humans and other living things.
4) **Environmental Justice** calls for universal protection from nuclear testing, extraction, production and disposal of toxic/hazardous wastes and poisons and nuclear testing that threaten the fundamental right to clean air, land, water, and food.
5) **Environmental Justice** affirms the fundamental right to political, economic, cultural and environmental self-determination of all peoples.

6) **Environmental Justice** demands the cessation of the production of all toxins, hazardous wastes, and radioactive materials, and that all past and current producers be held strictly accountable to the people for detoxification and the containment at the point of production.

7) **Environmental Justice** demands the right to participate as equal partners at every level of decision-making, including needs assessment, planning, implementation, enforcement and evaluation.

8) **Environmental Justice** affirms the right of all workers to a safe and healthy work environment without being forced to choose between an unsafe livelihood and unemployment. It also affirms the right of those who work at home to be free from environmental hazards.

9) **Environmental Justice** protects the right of victims of environmental injustice to receive full compensation and reparations for damages as well as quality health care.

10) **Environmental Justice** considers governmental acts of environmental injustice a violation of international law, the Universal Declaration On Human Rights, and the United Nations Convention on Genocide.

11) **Environmental Justice** must recognize a special legal and natural relationship of Native Peoples to the U.S. government through treaties, agreements, compacts, and covenants affirming sovereignty and self-determination.

12) **Environmental Justice** affirms the need for urban and rural ecological policies to clean up and rebuild our cities and rural areas in balance with nature, honoring the cultural integrity of all our communities, and provided fair access for all to the full range of resources.

13) **Environmental Justice** calls for the strict enforcement of principles of informed consent, and a halt to the testing of experimental reproductive and medical procedures and vaccinations on people of color.

14) **Environmental Justice** opposes the destructive operations of multi-national corporations.
15) **Environmental Justice** opposes military occupation, repression and exploitation of lands, peoples and cultures, and other life forms.
16) **Environmental Justice** calls for the education of present and future generations which emphasizes social and environmental issues, based on our experience and an appreciation of our diverse cultural perspectives.
17) **Environmental Justice** requires that we, as individuals, make personal and consumer choices to consume as little of Mother Earth's resources and to produce as little waste as possible; and make the conscious decision to challenge and reprioritize our lifestyles to ensure the health of the natural world for present and future generations.

Source: "Principles of Environmental Justice." 1991. Available from several sources, including Energy Justice Network. https://www.ejnet.org/ej/principles.html. Accessed on July 3, 2019.

Chemical Waste Management, Inc. v. Templet (1991)

In 1976, the state of Louisiana passed a law, "Hazardous wastes from foreign nations; findings; prohibitions," which prohibited the import of hazardous wastes from any foreign country into the states. A major provision of that law is reprinted below. In 1989, Chemical Waste Management, Inc. ("ChemWaste") informed the regional office of the Environmental Protection Agency (EPA) that they intended to begin importing hazardous wastes from Mexico, by land through the state of Texas, to its storage site in Louisiana. When the EPA noted and approved of Louisiana's law prohibiting such transport, ChemWaste brought suit against the secretary of the Louisiana Department of Environmental Quality, Paul H. Templet, arguing that the state law was unconstitutional. The company based its suit on the Commerce Clause (Article I, Section 8,

Clause 3) and Supremacy Clause (Article VI, Clause 2) of the U.S. Constitution. The court agreed with ChemWaste, pointing out that provisions of the federal Resource Conservation and Recovery Act of 1976 superseded the state law.

[The Louisiana law declared that:]

(1) The laws of the United States require testing, manifesting, and safe transportation of hazardous wastes to insure proper identification and handling from generation to ultimate disposal. These laws are not applicable to hazardous wastes generated in foreign nations until such wastes are actually in this country.

(2) The laws of foreign nations are inadequate to insure that hazardous wastes sought to be exported to the United States do not contain unknown or unauthorized pollutants and that such wastes are not released into the environment due to inadequate containment, labeling, or handling during transport.

(3) The only practical method for insuring that the environment and the health of the citizens of this state are not endangered by the importation of hazardous wastes generated in foreign nations is to prohibit the introduction or receipt of such wastes into this state for the purpose of treatment, storage, or disposal.

[The court concluded that:]

A review of the evidence presented in this case clearly shows that the effect of the statute was to restrict the flow of commerce from Texas to Louisiana. The Court must note that 40 CFR 271.7 requires that when a state seeks have [*sic*] its environmental program operate in lieu of the federal program, e.g., become an "authorized" state, the attorney general of that state must certify to the EPA that the state program is consistent with the federal program. In the certification opinion to the EPA, the Louisiana Attorney General addressed the EPA's concern as to the proper interpretation of the term "directly." The Louisiana Attorney General opined that the restriction only applied to hazardous waste imported directly into Louisiana from a foreign country. He also concluded it was not direct if the waste traveled through another state. Under this interpretation, the LDEQ has misapplied the statutes, causing them to be unconstitutional "as applied."

The federal RCRA program allows the importation of foreign hazardous waste. Consequently, foreign generated hazardous waste is legally imported to other states of the United States. The importer of the foreign generated hazardous waste must not only satisfy the federal requirements, but once the waste is legally in the United States, the transporter and disposer are subject to the same environmental laws that are applicable to domestic generated hazardous waste. Therefore, when Louisiana prohibits waste transported from Mexico to Louisiana through Texas, it implicates interstate commerce because the Texas environmental laws concerning transportation apply whether the waste is properly categorized as "directly" or "indirectly" imported.

. . .

IT IS ORDERED that Louisiana Revised Statutes 30:2190 and 30:2191 be and each is hereby declared unconstitutional.

Source: *Chemical Waste Management, Inc. v. Templet*, 770 F. Supp. 1142 (M.D. La. 1991). 1991.

Environmental Equity: Reducing Risks for All Communities (1992)

One of the most critical issues in discussions about solid waste disposal involves the question of environmental justice. Environmental justice refers to the remediation of the disproportionate impact of environmental pollution on communities of color, race, ethnic origin, income, or other factors. The movement arose in the early 1980s as evidence began to accumulate that corporations were constructing factories, waste disposal sites, and other facilities likely to cause environmental damage in communities that were least able to organize and act to protect themselves from such exposures. In 1991, Environmental Protection Agency administrator William K. Reilly appointed a commission to study environmental inequities in the United States based on race, income status, and other factors. The commission released its report on July 22, 1992, the main features of which are quoted below. The battle against environmental justice has continued, with somewhat limited success, over the years.

Summary of Findings

1. There are clear differences between racial groups in terms of disease and death rates. There are also limited data to explain the environmental contribution to these differences. In fact, there is a general lack of data on environmental health effects by race and income. For diseases that are known to have environmental causes, data are not typically dis-aggregated by race and socioeconomic group. The notable exception is lead poisoning: A significantly higher percentage of Black children compared to White children have unacceptably high blood lead levels.
2. Racial minority and low-income populations experience higher than average exposures to selected air pollutants, hazardous waste facilities, contaminated fish and agricultural pesticides in the workplace. Exposure does not always result in an immediate or acute health effect. High exposures, and the possibility of chronic effects, are nevertheless a clear cause for health concerns.
3. Environmental and health data are not routinely collected and analyzed by income and race. Nor are data routinely collected on health risks posed by multiple industrial facilities, cumulative and synergistic effects, or multiple and different pathways of exposure. Risk assessment and risk management procedures are not in themselves biased against certain income or racial groups. However,

risk assessment and risk management procedures can be improved to better take into account equity considerations.

4. Great opportunities exist for EPA and other government agencies to improve communication about environmental problems with members of low-income and racial minority groups. The language, format and distribution of written materials, media relations, and efforts in two-way communication all can be improved. In addition, EPA can broaden the spectrum of groups with which it interacts.
5. Since they have broad contact with affected communities, EPA's program and regional offices are well suited to address equity concerns. The potential exists for effective action by such offices to address disproportionate risks. These offices currently vary considerably in terms of how they address environmental equity issues. Case studies of EPA program and regional offices reveal that opportunities exist for addressing environmental equity issues and that there is a need for environmental equity awareness training. A number of EPA regional offices have initiated projects to address high risks in racial minority and low-income communities.
6. Native Americans are a unique racial group that has a special relationship with the federal government and distinct environmental problems. Tribes often lack the physical infrastructure, institutions, trained personnel and resources necessary to protect their members.

Summary of Recommendations

Although large gaps in data exist, the Workgroup believes that enough is known with sufficient certainty to make several recommendations to the Agency. These recommendations are also applicable to other public and private groups engaged in environmental protection activities. The job of achieving environmental equity is shared by everyone.

1. EPA should increase the priority that it gives to issues of environmental equity.
2. EPA should establish and maintain information which provides an objective basis for assessment of risks by income and race, beginning with the development of a research and data collection plan.
3. EPA should incorporate considerations of environmental equity into the risk assessment process. It should revise its risk assessment procedures to ensure, where practical and relevant, better characterization of risk across populations, communities or geographic areas. These revisions could be useful in determining whether there are any population groups at disproportionately high risk.
4. EPA should identify and target opportunities to reduce high concentrations of risk to specific population groups, employing approaches developed for geographic targeting.
5. EPA should, where appropriate, assess and consider the distribution of projected risk reduction in major rulemakings and Agency initiatives.
6. EPA should selectively review and revise its permit, grant, monitoring and enforcement procedures to address high concentrations of risk in racial minority and low-income communities. Since state and local governments have primary authority for many environmental programs, EPA should emphasize its concerns about environmental equity to them.
7. EPA should expand and improve the level and forms with which it communicates with racial minority and low-income communities and should increase efforts to involve them in environmental policy-making.
8. EPA should establish mechanisms, including a center of staff support, to ensure that environmental equity concerns are incorporated in its long-term planning and operations.

Source: "Environmental Equity: Reducing Risk for All Communities." 1992. Environmental Protection Agency. https://nepis.epa.gov/Exe/ZyPDF.cgi/40000JLA.PDF?Dockey=40000JLA.PDF. Accessed on June 29, 2019.

United States of America v. Asarco Incorporated (Consent Decree) (1998)

One of the primary mechanisms used by the U.S. Environmental Protection Agency in dealing with waste management problems that fall under the Resource Conservation and Recovery Act of 1976 or the Comprehensive Environmental Response, Compensation, and Liability Act of 1980 is a consent decree. One of the best-known instances in which a company agreed to follow this route in settling a case with the EPA occurred in 1998. On that occasion, ASARCO, a national mining and smelting company, agreed to invest more than $50 million to cleanup sites that had been contaminated by its production activities. A section of the very long agreement reached on this case is reprinted here.

Statement of Requirements

22. Through this Part of the Decree, ASARCO agrees to:

a. Identify, and provide to EPA on request, studies performed at, and data and information collected regarding environmental conditions at the Facility prior to the effective date of this Decree;
b. Review the effectiveness of any measures which would meet the definition of interim measures, performed at the Facility prior to the effective date of this Decree and submit this information to EPA in combination with the information required in Subparagraph a. above in a report entitled Current Conditions/Release Assessment;
c. Perform interim measures where possible and appropriate, at the Facility;

d. Perform a RCRA Facility Investigation ("RFI") to determine the full nature and extent of any and all releases of hazardous wastes and/or hazardous constituents at or from the Facility;

e. Perform a Corrective Measure Study ("CMS") to identify and evaluate alternatives which will prevent or mitigate the continuing migration of or future release of hazardous waste or hazardous constituents at and/or from the Facility, and to restore contaminated media to standards acceptable to EPA;

f. Implement all Corrective Measures ("CMI") chosen by EPA after review of the CMS and public input, which will be chosen because they best prevent or mitigate the continuing migration of or future release of hazardous waste or hazardous constituents at and/or from the Facility, and will result in the remediation of contaminated media in a manner protective of human health and the environment; and

g. Conduct such activities in accordance with existing regulations and guidance, including, the preamble to the regulations EPA proposed in 1990 to promulgate as 40 C.F.R. Part 264, Subpart S, the CAP and the ANPR; then applicable EPA regulations which supersede presently existing EPA regulations or guidance; or Montana regulations which have been incorporated into the federally authorized program which supersede existing EPA regulations or guidance.

. . .

XII. PENALTY FOR PAST VIOLATIONS

179. Within 30 days after the date of entry of this Decree, ASARCO shall pay to the United States a civil penalty in the amount of Three Million three Hundred and Eighty-Six thousand and One Hundred Dollars ($3,386,100), plus interest at the rate established by the Secretary of the Treasury pursuant to 31 U.S.C. § 3717, calculated from the date of lodging of this Decree until the date of payment.

XIII. STIPULATED PENALTIES

180. If ASARCO fails to submit any report required by Section XI (Reporting) on or before the specified due date, ASARCO shall pay a stipulated penalty of $1,000 per day. Provided, however, the time period during which stipulated penalties accrue for failure to submit a report shall not exceed the date the next such periodic report required by that particular part of the Decree is submitted if the later report contains all information required to be submitted in the first (and any subsequently missed) reports.

181. If ASARCO fails to comply with any other requirement of this decree, unless such noncompliance is excused pursuant to Part XV (Force Majeure), ASARCO shall pay stipulated civil penalties as follows:

Period of Failure to Comply Penalty

1st to 14th day $ 1,000/day per violation

15th to 30th day $ 2,000/day per violation

After 30 days $ 3,000/day per violation

Source: "*United States of America v. ASARCO Incorporated.*" 1998. Environmental Protection Agency. https://www.epa.gov/sites/production/files/2013-09/documents/asarcophs1cd.pdf. Accessed on July 1, 2019.

Not in My Backyard: Executive Order 12898 and Title VI as Tools for Achieving Environmental Justice (2003)

In 2003, the U.S. Commission on Civil Rights undertook a study to determine how well four government agencies—the Environmental Protection Agency, the U.S. Department of the Interior, the U.S. Department of Housing and Urban Development, and the U.S. Department of Transportation—have implemented Executive Order 12898 and Title VI of the 1964 Civil Rights Act. (Executive Order 12898, had been issued by President Bill Clinton in 1994 to focus on issues of environmental justice in the federal government. It contained instructions for specific actions to be taken to remedy the problem primarily in the four agencies listed here.) The commission concluded that these agencies had not made much progress in this regard. In its letter of transmittal to the President and the Congress, the commission made the following observation.

[S]ignificant problems and shortcomings remain. Federal agencies still have neither fully incorporated environmental justice into their core missions nor established accountability and performance outcomes for programs and activities. Moreover, a commitment to environmental justice is often lacking in agency leadership, communities are not yet full participants in environmental decision-making, and there is still inadequate scientific and technical literature on the relationship between environmental pollutants and human health status. Although poor communities and communities of color are becoming more skilled at using Title VI administrative processes to seek recourse and remedies, agencies seldom, if ever, revoke a permit or withhold money from the recipients of federal funding for violating Title VI. Strong administrative enforcement of Title VI is required in light of court decisions limiting access to judicial recourse and remedies under Title VI. Uncertainty about the use and effectiveness of Title VI in protecting the poor and communities of color is created by the absence of final investigative and recipient guidance by EPA. The agency was moving toward finalizing its Title VI guidance at the time the Commission report was drafted, and we look forward to its release. The other agencies, unlike EPA, lacked any comprehensive Title VI investigation and recipient guidance.

[Some of the conclusions offered by the Commission are as follows:]

A renewed effort by federal agencies to collect, analyze, and maintain data on risks and exposures be undertaken. Formal guidance on assessing cumulative risk should be created by federal agencies that considers the roles of social, economic, and behavioral factors when assessing risk. Guidance should include a presumption of adverse health risks when populations are exposed to multiple hazards from multiple sources. Federal agencies should disaggregate data on risks and exposures by race, ethnicity, gender, age, income, and geographic location if communities are to have the tools they need to defend environmental and human health and if agencies are to fulfill their obligations under Executive Order 12898 and Title VI. Federal agencies should require state and local zoning and land-use authorities, as a condition for receiving and continuing to receive federal funding, to incorporate and implement the principles of environmental justice into their zoning and land-use policies. The funding scheme for the Superfund program should be reviewed by Congress to ensure that the program is effectively funded and administered.

Source: U.S. Commission on Civil Rights. 2003. "Not in My Backyard: Executive Order 12898 and Title VI as Tools for Achieving Environmental Justice." U.S. Commission on Civil Rights. iii, 27–28. https://www.usccr.gov/pubs/envjust/ej0104.pdf. Accessed on June 29, 2019.

Electronic Waste Recycling Act (2003)

In addition to federal laws dealing with waste management, several states have adopted legislation extending such laws in one or another area of special interest. The California legislature went down this route in 2003 when it passed the Electronic Waste Recycling Act (EWRA) of that year. A key portion of that act is as follows:

42461. The Legislature finds and declares all of the following:

(a) The purpose of this chapter is to enact a comprehensive and innovative system for the reuse, recycling, and proper and legal disposal of covered electronic devices, and to provide incentives to design electronic devices that are less toxic, more recyclable, and that use recycled materials.

(b) It is the further purpose of this chapter to enact a law that establishes a program that is cost free and convenient for consumers and the public to return, recycle, and ensure the safe and environmentally-sound disposal of covered electronic devices.

(c) It is the intent of the Legislature that the cost associated with the handling, recycling, and disposal of covered electronic devices is the responsibility of the producers and consumers of covered electronic devices, and not local government or their service providers, state government, or taxpayers.

(d) In order to reduce the likelihood of illegal disposal of these hazardous materials, it is the intent of this chapter to ensure that any cost associated with the proper management of covered electronic devices be internalized by the producers and consumers of covered electronic devices at or before the point of purchase, and not at the point of discard.

(e) Manufacturers of covered electronic devices, in working to achieve the goals and objectives of this chapter, should have the flexibility to partner with each other and with those public sector entities and business enterprises that currently provide collection and processing services to develop and promote a safe and effective covered electronic device recycling system for California.

(f) The producers of electronic products, components, and devices should reduce and, to the extent feasible, ultimately phase out the use of hazardous materials in those products.

(g) Electronic products, components, and devices, to the greatest extent feasible, should be designed for extended life, repair, and reuse.

(h) The purpose of the Hazardous Electronic Waste Recycling Act is to provide sufficient funding for the safe, cost-free, and convenient collection and recycling of 100 percent of the covered electronic waste discarded or offered for recycling in the state, to eliminate electronic waste stockpiles and legacy devices by December 31, 2007, to end the illegal disposal of covered electronic devices, to establish manufacturer responsibility for reporting to the board on the manufacturer's efforts to phase out hazardous materials in electronic devices and increase the use of recycled materials, and to ensure that electronic devices sold in the state do not violate the regulations adopted by the Department of Toxic Substances Control pursuant to Section 25214.10 of the Health and Safety Code.

[The term "covered electronic device" is defined in the act as follows:]

(f)(1) "Covered electronic device" means a cathode ray tube, cathode ray tube device, flat panel screen, or any other similar video display device with a screen size that is greater than four inches in size measured diagonally and which the department determines, when discarded or disposed, would be a hazardous waste pursuant to Chapter 6.5 (commencing with Section 25100) of Division 20 of the Health and Safety Code.

(2) "Covered electronic device" does not include an automobile or a large piece of commercial or industrial equipment, including, but not limited to, commercial medical equipment, that contains a cathode ray tube, cathode ray tube device, flat panel screen, or other similar video display device that is contained within, and is not separate from, the larger piece of industrial or commercial equipment.

Source: SB 20, Sher, Chapter 526, Statutes of 2003. September 25, 2003. http://www.leginfo.ca.gov/cgi-bin/statquery. Accessed on June 16, 2019.

Municipal Solid Waste Landfills (2018)

The U.S. Congress and various administrative agencies of the government have, over the years, established very specific and extended requirements for the establishment of a solid waste landfill. Those requirements take up 50 pages of the U.S. Code of Federal Regulations. The selection below, from the website of the Environmental Protection Agency, summarizes the most important features of these regulations.

What Is a Municipal Solid Waste Landfill?

A municipal solid waste landfill (MSWLF) is a discrete area of land or excavation that receives household waste. A MSWLF may also receive other types of nonhazardous wastes, such as commercial solid waste, nonhazardous sludge, conditionally exempt small quantity generator waste, and industrial nonhazardous solid waste. In 2009, there were approximately 1,908 MSWLFs in the continental United States all managed by the states where they are located.

Nonhazardous solid waste is regulated under Subtitle D of RCRA. States play a lead role in ensuring the federal criteria for operating municipal solid waste and industrial waste landfills regulations are met, and they may set more stringent requirements. In absence of an approved state program, the federal requirements must be met by waste facilities. The revised criteria in Title 40 of the Code of Federal Regulations (CFR) part 258 addresses seven major aspects of MSWLFs, which include the following:

Location restrictions—ensure that landfills are built in suitable geological areas away from faults, wetlands, flood plains or other restricted areas. Composite liners requirements—include a flexible membrane (i.e., geo-membrane) overlaying two feet of compacted clay soil lining the bottom and sides of the landfill. They are used to protect groundwater and the underlying soil from leachate releases. Leachate collection and removal systems—sit on top of the composite liner and removes leachate from the landfill for treatment and disposal. Operating practices—include compacting and covering waste frequently with several inches of soil. These practices help reduce odor, control litter, insects, and rodents, and protect public health. Groundwater monitoring requirements—requires testing groundwater wells to determine whether waste materials have escaped from the landfill. Closure and post-closure care requirements—include covering landfills and providing long-term care of closed landfills. Corrective action provisions—control and clean up landfill releases and achieves groundwater protection standards. Financial assurance—provides funding for environmental protection during and after landfill closure (i.e., closure and post-closure care).

Some materials may be banned from disposal in MSWLFs, including common household items like paints, cleaners/chemicals, motor oil, batteries and pesticides. Leftover portions of these products are called household hazardous waste. These products, if mishandled, can be dangerous to your health and the environment. Many MSWLFs have a household hazardous waste drop-off station for these materials.

MSWLFs can also receive household appliances (i.e. white goods) that are no longer needed. Many of these appliances, such as refrigerators or window air conditioners, rely on ozone-depleting refrigerants and their substitutes. MSWLFs follow the federal disposal procedures for household appliances that use refrigerants. EPA has general information on how refrigerants can damage the ozone layer and consumer information on the specifics for disposing of these appliances.

Municipal Solid Waste Transfer Stations

Waste transfer stations are facilities where municipal solid waste (MSW) is unloaded from collection vehicles. The MSW is briefly held while it is reloaded onto larger long-distance transport vehicles (e.g. trains, trucks, barges) for shipment to landfills or other treatment or disposal facilities. Communities can save money on the labor and operating costs of transporting the waste to a distant disposal site by combining the loads of several individual waste collection trucks into a single shipment.

They can also reduce the total number of trips traveling to and from the disposal site. Although waste transfer stations help reduce the impacts of trucks traveling to and from the disposal site, they can cause an increase in traffic in the immediate area where they are located. If not properly sited, designed and operated they can cause problems for residents living near them.

A Regulatory Strategy for Siting and Operating Waste Transfer Stations provides information about waster transfer stations and the actions EPA has taken to address this issue.

Source: "Municipal Solid Waste Landfills." 2018. Environmental Protection Agencies. https://www.epa.gov/landfills/municipal-solid-waste-landfills. Accessed on July 5, 2019. The full description of requirements can be found at https://www.govinfo.gov/content/pkg/CFR-2011-title40-vol25/pdf/CFR-2011-title40-vol25-part258.pdf.

Negotiating Superfund Settlements (2019)

Identifying sites that might qualify for treatment under the EPA's Superfund program and then arranging for the cleanup of such sites is often a complex and expensive task. In many instances, a company will disagree with the EPA in its assessment of the status of a possible hazard waste site. Such disagreements occasionally go to court, but, in general, the EPA prefers to come to a pretrial settlement with the prospective defendant in such an action. The following statement describes the most common ways in which the EPA and possible defendants in a Superfund action can reach an agreement over possible actions. Links to related pages are omitted here.

EPA prefers to reach an agreement with a potentially responsible party (PRP) to clean up a Superfund site instead of issuing an order or paying for it and recovering the cleanup costs later.

Superfund settlement agreements must be:

In the public interest, and Consistent with the National Contingency Plan.

A special notice letter invites a PRP to enter into good faith negotiations and gives the PRP 60 days to provide EPA with a good faith offer to do site work or pay for cleanup. If the PRP provides a good faith offer, there is generally another 60 days for negotiation. If the PRP does not submit a good faith offer at the end of 60 days, EPA may start the cleanup work or issue a unilateral administrative order, requiring the PRPs to do the work.

Types of Superfund Settlement Agreements with PRPs

A Superfund cleanup agreement is written in the form of an administrative settlement agreement and order on consent (ASAOC) or a judicial consent decree (CD). Negotiations are based on model settlement agreements, which are usually modified to fit the circumstances at a particular site. Model settlement documents are available from the Cleanup Enforcement Model Language and Sample Documents database. Additional information on Superfund settlements is available from the Superfund enforcement cleanup policy and guidance database.

Administrative Settlement Agreement and Order on Consent

An ASAOC is a legal document that formalizes an agreement between EPA and one or more PRPs, to address some or all of the parties' responsibility for a site. Administrative orders on consent do not require approval by the court.

EPA uses ASAOCs for removal activity (short-term cleanup), remedial investigation and feasibility studies, and remedy design work.

EPA also uses ASAOCs for cost recovery when the payments are made as part of an agreement for work and for *de minimis* cashout payments.

Judicial Consent Decrees

A consent decree (CD) is a legal agreement entered into by the United States (through EPA and the Department of Justice) and PRPs. CDs are lodged with a court.

Consent decrees are the only settlement type that EPA can use for the final cleanup phase (remedial action) at a Superfund site. EPA also uses CDs to recover cleanup costs in cost recovery and cashout settlements and on rare occasions to perform removal work or remedial investigations/feasibility studies.

A consent decree is final when it is approved and entered by a U.S. district court.

Types of Settlement Agreements

- Administrative Agreement
- Agreement for "Work"
- Cost Recovery Agreement
- "Cashout" Agreement

Administrative Agreement

An administrative agreement is a legal document that formalizes an agreement between EPA and one or more PRPs to reimburse EPA for costs already incurred (cost recovery) or for costs to be incurred (cashout) at a Superfund site. (Cashout settlements generally include payments for both past and future costs, but always include a future cost component.)

An Administrative agreement does not require approval by the court. All types of payment agreements that do not include performance of work are generally written as an administrative agreement.

Agreement for "Work"

EPA prefers that PRPs do the work of investigating, cleaning up, and maintaining the cleanup of Superfund sites. EPA negotiates an agreement (in the form of an ASAOC or CD) with the PRP that outlines the work that is to be done.

The term "work agreement" is used to cover a variety of agreements that involve the PRP doing the work (versus EPA doing the work). The most common agreements are for:

> site investigation (remedial investigation and feasibility study), short-term clean up (removal action), and long-term cleanup (remedial design/remedial action).

Cost Recovery Agreement

When EPA performs investigations or cleanup work, it can recover its costs from PRPs through a cost recovery agreement.

When an agreement only addresses reimbursing EPA costs, it is referred to as a Cost Recovery Agreement and takes the form of an Administrative Agreement.

Administrative orders on consent for work may include a provision for the PRP to reimburse EPA for past work costs and will include a provision for the PRP to pay EPA's future costs in overseeing the PRP's work (considered "cost recovery" because such costs are billed to PRPs after the costs are incurred by EPA).

"Cashout" Agreement

There are a few situations when it is more appropriate for PRPs not to be involved in performing work at a site. In such cases, EPA may negotiate a "cashout" agreement with the PRP, where the PRP pays an appropriate amount of estimated site costs in advance of the work being done. That money will generally be used to help pay for the cleanup.

Third-Party Agreements

- Bona Fide Prospective Purchaser Agreement
- Prospective Purchaser Agreement
- Other Third-Party Agreement

Bona Fide Prospective Purchaser Agreement

The activities of most bona fide prospective purchasers (BFPPs) will not require liability protection beyond what is provided by the self-implementing BFPP protection in CERCLA § 107(r). However, if a BFPP wants to perform cleanup work at a contaminated site of federal interest that exceeds the BFPP's statutory requirements (e.g., reasonable steps), a work agreement may be used to address potential liability concerns associated with that cleanup work.

Prospective Purchaser Agreement

In limited circumstances, a prospective purchaser agreement (PPA) may be appropriate for a party who does not qualify for the statutory BFPP protection. Prospective purchaser agreements, similar to BFPP agreements, provide liability protections in exchange for cleanup work at a site of federal interest.

Other Third-Party Agreements

EPA may consider entering into a windfall lien agreement when a BFPP wants to satisfy any existing or potential windfall lien before or close to the time of acquisition.

For contiguous property owners (CPOs) who have been sued under CERCLA, or can demonstrate a real and substantial threat of such litigation, especially where the EPA has conducted a response action on the neighboring contaminated property or the contiguous property owner's property, EPA will consider whether a CPO agreement is appropriate to provide the owner with cost recovery or contribution protection from PRPs at the site.

EPA also developed a Good Samaritan settlement agreement to provide a federal CERCLA covenant not to sue and contribution protection, when appropriate, to certain parties who volunteer to perform cleanup work at an orphan hard rock mine site.

Long-Term Stewardship/Institutional Controls in Superfund Settlements

Superfund settlement agreements usually include long-term stewardship obligations to maintain the cleanup. Long-term stewardship activities typically include physical and legal controls to prevent exposure to contamination left in place at a site. For example, a groundwater cleanup may involve operating a treatment system for 30 or more years. At such a site, a legal control such as a groundwater use regulation may be used to meet the long-term stewardship obligation.

The legal controls are generally referred to as "institutional controls" (ICs). EPA uses ICs to help:

- minimize human exposure to contamination, and
- protect structures and systems that are part of the cleanup (such as groundwater monitoring systems and landfill covers).

Institutional controls are considered part of the remedy for the site. How ICs are enforced depends on the nature of the control and how it is initiated (e.g., through a local ordinance; in an enforceable agreement).

EPA continually assesses the status of ICs at Superfund sites and gathers IC data from Superfund sites where construction of the remedy is complete.

Examples of ICs include:

- conservation
- covenants deed or hazard notices
- environmental easements
- restrictions on groundwater use
- special building permit requirements
- state or local ordinances state registries of contaminated properties
- well drilling prohibitions
- zoning restrictions

Further information on institutional controls is available on the Institutional Controls webpage and from the institutional controls category within the Superfund enforcement policy and guidance database.

Source: "Negotiating Superfund Settlements." 2019. Environmental Protection Agency. https://www.epa.gov/enforcement/negotiating-superfund-settlements. Accessed on July 1, 2019.

Waste management has been an issue with which human societies have had to deal since urban areas first developed more than 5,000 years ago. Discussions as to what exactly the nature of waste problems is and how they are to be dealt with have been the subject of countless written documents. As chapter 1 has illustrated, some of those documents can be traced to the ancient Mesopotamian civilizations dating to before 3000 BCE. The volume of books, articles, reports, and internet web pages dealing with waste management is, therefore, enormous. No single resource can include more than a small fraction of those writings. This chapter contains samples of some of the best-known and most useful books, articles, and other documents on the topic of waste management. In some cases, a document may be available in more than one format, a journal article and web page, for example—a fact that is so indicated in the section where it is to be found in this bibliography. Readers should also be aware of the wealth of resource ideas contained in the notes at the end of chapters 1 and 2.

Books

Abbing, Michael Roscam. 2019. *Plastic Soup: An Atlas of Ocean Pollution*. Washington, DC: Island Press.

Improperly discarded electronic equipment, known as e-waste, is of growing environmental concern. Items such as computers, televisions, and cell phones may leach toxins such as cadmium, lead, and mercury into freshwater ecosystems. (Shutterstock)

This book is a heavily illustrated discussion of all aspects of the issue of plastic waste pollution in the world's oceans.

Al-Salem, Sultan. 2019. *Plastics to Energy: Fuel, Chemicals, and Sustainability Implications*. Kidlington, Oxford, UK: William Andrew.

This book discusses developments in end-of-life procedures for treating plastic wastes. It discusses features that will be of interest to engineers, material scientists, waste management administrators, and the general public. It also reviews the latest trends in the practical applications of these procedures.

Ashton, Rosemary. 2018. *One Hot Summer: Dickens, Darwin, Disraeli, and the Great Stink of 1858*. New Haven: Yale University Press.

The author provides a detailed discussion of the hot summer months of 1858 and the public health problems that occurred because of that event, with special attention to individual persons of note, public health consequences, and legislative actions resulting from the period.

Barnes, David S. 2018. *The Great Stink of Paris and the Nineteenth-Century Struggle against Filth and Germs*. Baltimore: Johns Hopkins University Press.

The very hot summer of 1880 produced striking environmental effects in Paris and, as it developed, serious public health issues in the city. The author explains how the event changed people's view of the cause and nature of disease.

Banidickson, Jamie. 2007. *The Culture of Flushing: A Social and Legal History of Sewage*. Vancouver: University of British Columbia Press.

The evolution of toilets and sanitation systems provides an insight into the attitudes about and method of waste management in Canada, the United Kingdom, and the

United States. This book reviews the main feature of that story.

Bullard, Robert D. 2018. *Dumping in Dixie Race, Class, and Environmental Quality*. 3rd ed. New York: Routledge.

Bullard, often referred to as the father of environmental justice, has now twice revised his 2000 text, generally regarded as one of the most important texts in the field of environmental justice.

Cabaniss, Amy D., ed. 2018. *Handbook on Household Hazardous Waste*. Lanham, MD: Bernan Press.

After opening with two chapters on the definitions of household hazardous wastes and the history of their management, six additional chapters deal with collection, educational programs, and the future of household hazardous waste collection programs.

Extended Producer Responsibility: A Guidance Manual for Governments. 2001. Paris: Organisation for Economic Co-operation and Development.

One approach recently developed for dealing with waste management has been called extended producer responsibility (EPR). The system involves making it the manufacturer's responsibility to take some role in determining the eventual fate of products they make, as in building in recycling or reuse options for those products. This book is an early effort to set out standards and methods for such programs. Also see, for example, Huang, Atasu, and Toktay 2015, under "Internet" below.

Galanakis, Charis M., ed. 2019. *Saving Food: Production, Supply Chain, Food Waste and Food Consumption*. London: Academic Press.

The thirteen essays in this book deal with all aspects of food loss and food waste. The majority of articles focus on advances in technology for reducing food wastes, such as improving agricultural practices, food processing, food storage technologies, developing better packaging technologies, and finding more efficient ways to transport food products.

Ghosh, Dhrubajyoti. 2017. *The Trash Diggers*. New Delhi: Oxford University Press.

This book introduces the trash pickers of Dhapa, India, and the lifestyle they pursue in manually sorting and removing components from trash heaps located in their community. It describes a way of life that is necessary for the communities involved but that can be devastating to the health of the pickers.

Goel, Sudha, ed. 2017. *Advances in Solid and Hazardous Waste Management*. New Delhi: Springer.

This book provides more than a dozen essays on specific advances that have been made in the field of waste management technology. Some topics relate to the use of remote sensing and GIS systems in waste management, leaching in pond ash, degradation of plastics, and special issues related to electronic wastes.

Guidelines for Framework Legislation for Integrated Waste Management. 2016. United Nations Environment Programme. https://wedocs.unep.org/bitstream/handle/20.500.11822/22098/UNEP%20Guidelines%20for%20Framework%20Legsilation%20for%20Integrated%20Waste%20Managment.pdf. Accessed on July 8, 2019.

This very detailed document provides guidelines for government officials on the theory and practice of dealing with waste products. It includes not only technical information on the methods available for solid waste management but also economic, political, social, legal, and other issues in solid waste management programs.

Hardt, Karen. 2018. *Solid Waste Management*. Forest Hills, NY: Callisto Reference.

This textbook provides a general overview of and introduction to the topic of solid waste management.

Harrison, R. M., and R. E. Hester. 2018. *Plastics and the Environment*. London: Royal Society of Chemistry.

This volume is of special interest because, in addition to the usual review of marine plastic pollution, it covers topics such as microplastics, nanoplastics, plasticizers, and plastic additives and their effects on human health.

Jafarinejad, Shahryar. 2017. *Petroleum Waste Treatment and Pollution Control*. Oxford, UK; Cambridge, MA: Butterworth-Heinemann.

This book provides an excellent introduction to the history and current status of the petroleum industry and the wastes with which it is associated. It covers solid, liquid, and gaseous wastes; methods that have been developed to treat those wastes; and governmental regulations dealing with waste issues.

Lindenlauf, Astrid. 2001. *Waste Management in Ancient Greece from the Homeric to the Classical Period: Concepts and Practices of Waste, Dirt, Recycling and Disposal*. Doctoral dissertation. University of London.

This difficult-to-find document provides an exhaustive and insightful overview of the treatment of solid wastes by the ancient Greeks.

Macaskie, Lynne E., D. J. Sapsford, and Will M. Mayes, eds. 2019. *Resource Recovery from Wastes: Towards a Circular Economy*. London: Royal Society of Chemistry.

The chapters in this book explore some specific examples of the use of the circular economy in helping to solve the world's waste management problems. Some examples include "Use of Biotechnology for Conversion of Lignocellulosic Waste into Biogas and Renewable Chemicals," "Towards an Integrated Hydrogen and Bioenergy Biorefinery," "Resource Recovery Using Microbial Electrochemical Technologies," and "In Situ Recovery of Resources from Waste Repositories."

McCallum, Will. 2018. *How to Give up Plastic: A Guide to Changing the World, One Plastic Bottle at a Time.* New York: Penguin Books.

This book is a very practical guide as to how it is possible to eliminate the use of plastic objects in every part of a person's life.

Mitchell, Piers D. 2016. *Sanitation, Latrines and Intestinal Parasites in Past Populations.* London; New York: Routledge.

This book provides a very interesting summary of sanitary practices in prehistoric societies, ancient Mesopotamia, ancient Greece and Rome, the Middle Ages, and more recent ages.

Muralikrishna, Iyyanki V., and Valli Manickam. 2017. "Hazardous Waste Management." *Environmental Management.* Oxford, UK: Butterworth-Heinemann.

This chapter covers the major features of hazardous waste management, such as effects on human health; sampling and analysis methods; treatment, storage, and disposal; design and operation of landfills; and safety and occupation hygiene.

Niaounakis, Michael. 2017. *Management of Marine Plastic Debris: Prevention, Recycling, and Waste Management.* Oxford, UK; Cambridge, MA: Elsevier.

This book provides an excellent overview of most issues relating to the management of plastic wastes in the oceans, with chapters on environmental, social, and economic impacts; degradation of plastics in the marine environment; prevention and mitigation, and regulatory framework.

Pellow, David Naguib. 2004. *Garbage Wars: The Struggle for Environmental Justice in Chicago*. Cambridge, MA: MIT Press.
The author reviews the long-standing and bitter battle in Chicago over the siting of solid waste disposal sites in largely African American and low-income neighborhoods.

Prasad, Majeti Narasimha Vara, and Meththika Vithanage, eds. 2019. *Electronic Waste Management and Treatment Technology*. Amsterdam: Butterworth-Heinemann.
The essays in this volume focus on a variety of topics relating to electronic wastes, including statistics for the problem worldwide, environmental management of e-wastes, biorecovery of useful materials from e-wastes, and examples of e-wastes in Australia, Brazil, and other parts of the world.

Rada, Elena Cristina, ed. 2017. *Waste Management and Valorization: Alternative Technologies*. Oakville, ON: Apple Academic Press,
The term *valorization of wastes* references to any process by which wastes are treated in such a way as to produce new materials with economic value. The essays in this book provide introductions to a number of aspects of this topic, including aerobic waste management systems, recycling of incinerator ash, processing of plastic wastes, and the conversion of glass wastes to high-strength mortars.

Rathoure, Ashok K., ed. 2020. *Zero Waste: Management Practices for Environmental Sustainability*. Boca Raton, FL: CRC Press.
The chapters in this book cover several fundamental practices and suggestions for the development of zero waste programs. They include biomedical waste treatment and disposal, construction waste, plastic waste management

practices, treatment and disposal of hazardous waste, fugitive dust control in industry, debris management, wetland and biodiversity spot, fly ash management, industrial waste management systems, hazardous waste management and public, domestic waste management, and corporate responsibility for environmental protection.

Reinhardt, Peter A., and Judith G. Gordon. 2018. *Infectious and Medical Waste Management.* Boca Raton, FL: CRC Press.

Intended primarily for specialists in the field, this text provides useful information understandable to the layperson on all aspects of medical waste management.

Rissanen, Timo, and Holly McQuillan. 2018. *Zero Waste Fashion Design.* London: Bloomsbury.

Experts in many occupations are now exploring ways in which the concept of zero waste can be applied to their own fields. This book on fashion design, for example, contains chapters on a history of the practice in the field of fashion design, pattern cutting, zero waste garments, and fashion design and CAD (computer-assisted design).

Saxena, Gaurav, and Ram Naresh Bharagava, eds. 2020. *Bioremediation of Industrial Waste for Environmental Safety.* Volume I: Industrial Waste and Its Management; Volume II: Biological Agents and Methods for Industrial Waste Management. Singapore: Springer.

These two volumes provide a general overview of the nature of industrial wastes and methods by which they can be treated. The second volume concentrates on the use of biological agents for industrial waste remediation.

Sengupta, Sukalyan. 2018. *Hazardous Waste Management.* Volume I. New York: Momentum Press.

This book discusses all aspects of hazard waste management, from its recent history to its regulatory framework to related environmental chemistry to the fate and transport of contaminants to toxicology and risk assessment.

Shahnawaz, Mohd., Manisha K. Sangale, and Avinhash B. Ade. 2019. *Bioremediation Technology for Plastic Waste*. Singapore: Springer Verlag.

This excellent resource provides a thorough discussion of all aspects of plastic waste, including microplastics, plastic waste disposal, case studies of plastic waste degradation, the role of bacteria in bioremediation of plastics, and social awareness of plastic waste issues.

Singh, Ram Lakhan, and Rajat Pratap Singh, eds. 2019. *Advances in Biological Treatment of Industrial Waste Water and Their Recycling for a Sustainable Future*. Singapore: Springer.

The essays in this book deal with the treatment and recycling of wastewater from a variety of industries, such as pulp and paper, tanneries, the dairy industry, distilleries, wineries, sugar mills, and the textile and pharmaceutical industries.

Singhania, Reeta Rani, et al., eds. 2018. *Waste to Wealth*. Singapore: Springer.

One of the great challenges of solid waste management is to find ways in which wastes can not only be recycled, incinerated, composted, or treated in some efficient way but also to do so in a manner that will provide benefits to companies and societies. The twenty-one papers in this book describe some of the ways in which this objective has been achieved.

Solid Waste Management for Northern and Remote Communities: Planning and Technical Guidance Document. 2017. Gatineau QC: Environment and Climate Change Canada. http://publications.gc.ca/collections/collection_2017/eccc/En14-263-2016-eng.pdf. Accessed on July 8, 2019.

Communities that are geographically isolated and/or that face special weather and climate conditions may encounter solid waste management problems significantly different from those in more temperate and populated regions. This document was prepared by the Canadian government to provide assistance to such challenged communities in the northern territories and other remote regions of the country.

Sosna, Daniel, and Lenka Brunclíková, eds. 2017. *Archaeologies of Waste: Encounters with the Unwanted*. Oxford, UK; Philadelphia: Oxbow Books.

This fascinating book contains chapters on the nature of waste disposal systems in early culture and how those systems seem to have had social meaning to residents of the time.

Stahel, Walter R. 2019. *The Circular Economy: A User's Guide*. Abingdon, Oxon; New York: Routledge.

The author of this book is one of the pioneers of the circular economy movement. In it, he explains the philosophy behind the movement along with some practical applications of it.

Szaky, Tom, et al. 2019. *The Future of Packaging: From Linear to Circular*. Oakland, CA: Berrett-Koehler Publishers.

The fifteen essays in this book focus on the problem of waste products from packaging materials. Some topics include "The State of the Recycling Industry," "Who Is Responsible for Recycling Packing?," "Designing Packages for the Simple Recycler," "The Myth of Biodegradability," "The Forgotten Ones: Pre-Consumer Waste," and "Value for Business in the Circular Economy."

Taherzadeh, Mohammad, et al. 2019. *Sustainable Resource Recovery and Zero Waste Approaches*. St. Louis: Elsevier.

The essays in this book discuss a variety of topics related to the zero waste movement, in which wastes are converted into useful products and materials with essentially no final waste in the process. Some articles include "Agricultural, Industrial, Municipal and Forestry Wastes: An Overview,"

"Life Cycle Assessments of Waste Management," "Waste Refinery," "Sustainable Management of Solid Waste," "Source Separation of Household Waste: Technology and Social Aspects," "Composting of Wastes," "Vermicomposting of Wastes," and "Biogas from Wastes: Processes and Applications."

Wagner, Martin, and Scott Lambert, eds. 2018. *Freshwater Microplastics: Emerging Environmental Contaminants?* Cham, Switzerland: Springer. https://www.researchgate.net/publication/321225506_Freshwater_Microplastics_Emerging_Environmental_Contaminants. Accessed on July 27, 2019.

The study of freshwater plastic pollution is probably somewhat less popular than is the study of comparable marine pollution. The articles in this book comprise thirteen topics of special relevance to this issue, including sources and fates of microplastics in urban environments and in inland water in Asia, interactions of microplastics and freshwater biota, microplastic-associated biofilms, and issues of regulation and management of microplastic pollution.

Washington, Harriet A. 2019. *A Terrible Thing to Waste: Environmental Racism and Its Assault on the American Mind.* New York: Little, Brown Spark.

One of the most important recent books on environmental justice, this work argues that the disproportionate exposure to waste that people of color and low-income communities experience has direct effects on the mental development of those individuals. She discusses several specific instances in which exposure to waste products in the United States has had just that effect.

Wong, Jonathan W. C., et al. 2016. *Sustainable Solid Waste Management.* Reston, VA: American Society of Civil Engineers.

This book provides a somewhat more technical introduction to the topic of solid waste management, with chapters on public policies, recycling and minimizing disposal amounts, technologies for treating and recycling solid waste, development and maintenance of engineered landfills, landfill mining; and legal issues affecting the waste management industry.

Wong, Jonathan W. C., Tyagi, Rajeshwar D., and Ashok Pandey, eds. 2017. *Current Developments in Biotechnology and Bioengineering. Solid Waste Management*. Amsterdam; Cambridge, MA: Elsevier.

The twenty chapters in this book discuss some of the most recent developments in the conversion of solid wastes to useful materials, such as bioplastics from solid wastes, products from sewage sludge, biopesticides from solid wastes, conversion of organic materials to energy, ethanol-related products from agricultural wastes, recovery of useful metals from leachates, and biological treatments of hazardous wastes.

Articles

Several journals are entirely devoted to the topic of waste management or publish numerous articles on the topic. These journals include the following:

Advances in Recycling & Waste Management. ISSN: 2475–7675 (online only)

Compost Science and Utilization. ISSN: 1065–657X (print); 2326–2397 (online)

Critical Reviews in Environmental Science and Technology. ISSN: 1064–3389 (online)

Environmental Science and Pollution Research. ISSN: 0944–1344 (print); 1614–7499 (online)

International Journal of Environment and Waste Management. ISSN: 1478–9876 (print); 1478–9868 (online)

Journal of Environmental Management. ISSN: 1095–8630 (print) 0301–4797 (online)

Journal of Hazardous Materials. ISSN: 0304–3894

Journal of Material Cycles and Waste Management. ISSN: 1438–4957 (print) 1611–8227 (online)

Recycling. ISSN 2313–4321 (online only)

Resources, Conservation and Recycling. ISSN: 0921–3449 (online only)

Tribal Waste Journal (EPA publication). EPA530-N-09–002

Waste and Resource Management. ISSN: 1747–6526

Waste Atlas. ISSN: 2241–2484

Waste Management (Amsterdam). ISSN: 0956–053X

Waste Management & Research. ISSN: 0734–242X (print) 1096–3669 (online)

Waste Management World (online only)

Worldwide Waste. ISSN: 2399–7117 (online only)

Abbasi, S. A. 2018. "The Myth and the Reality of Energy Recovery from Municipal Solid Waste." *Energy, Sustainability and Society* 8(1): 1–15. https://energsustainsoc.biomedcentral.com/track/pdf/10.1186/s13705-018-0175-y. Accessed on July 9, 2019.

After pointing out the serious waste management problem faced by the world, the author suggests that at least some technologies developed for dealing with the problem are inadequate or insufficient. He argues that "what is technically feasible is economically unfeasible. And what is economically feasible . . . is exceedingly harmful to the environment and the human health."

Al Ansari, Mohammed Saleh. 2012. "Improving Solid Waste Management in Gulf Co-operation Council States: Developing Integrated Plans to Achieve Reduction in Greenhouse Gases." *Modern Applied Science* 6(2): 60–68. https://pdfs.semanticscholar.org/cdf0/b212a043f29c784a30284d57fd2bbcdde791.pdf. Accessed on July 6, 2019.

The decomposition of components of a sanitary landfill produces greenhouse gases that contribute to global warming. This paper explores the nature of that problem and some of the steps that are being taken by a group of Middle East countries to deal with the situation.

Arancon, Rick Arneil D., et al. 2013. "Advances on Waste Valorization: New Horizons for a More Sustainable Society." *Energy Science and Engineering* 1(2): 53–71. https://onlinelibrary.wiley.com/doi/full/10.1002/ese3.9. Accessed on July 24, 2019.

This excellent review article provides an overview of progress that has been made in the field of valorization of wastes, the conversion of waste materials to more valuable materials and products.

Bond, Tom, et al. 2013. Ancient Water and Sanitation Systems—Applicability for the Contemporary Urban Developing World. *Water Science and Technology* 67(5): 935–941. https://www.researchgate.net/publication/235648775_Ancient_water_and_sanitation_systems_-_applicability_for_the_contemporary_urban_developing_world. Accessed on July 8, 2019.

The authors provide a good review of sanitation systems known in ancient times and argue that the elements of those systems can be applied to sanitation systems at many points in developing countries around the world today.

Cánovas, Carlos Ruiz, et al. 2018. "Valorization of Wastes from the Fertilizer Industry: Current Status and Future Trends." *Journal of Cleaner Production* 174: 678–690.

Valorization, the conversion of wastes to useful industrial and commercial products, is the topic of a considerable amount of research today. This article provides an example of the kind of research being conducted and the results obtained from such research.

Coomans, Janna. 2018. "The King of Dirt: Public Health and Sanitation in Late Medieval Ghent." *Urban History* 46(1): 82–105.

https://www.cambridge.org/core/services/aop-cambridge-core/content/view/B730B9F9E0B88C4D88B9607BF36A73EA/S096392681800024Xa.pdf/div-class-title-the-king-of-dirt-public-health-and-sanitation-in-late-medieval-ghent-div.pdf. Accessed on July 28, 2019.

The author introduces the concept of *coninc der ribauden* as an administrative official responsible for, among other things, waste management in 14th- and 15th-century Ghent. The article provides an illuminating look at the way waste management was practiced during the Middle Ages.

Cooper, Daniel R., and Timothy G. Gutowski. 2017. "The Environmental Impacts of Reuse: A Review." *Journal of Industrial Ecology* 21(1): 38–56.

Research has shown that repairing and otherwise finding ways to reuse a "worn-out" product is less costly than remaking that product from scratch. Relatively little research has been done, however, on the possible environmental impacts of this approach to waste management, a topic about which the article is concerned.

Copeland, Claudia. 2016. "Ocean Dumping Act: A Summary of the Law." Congressional Research Service. https://fas.org/sgp/crs/misc/RS20028.pdf. Accessed on July 27, 2019.

This review articles provides information about almost any aspect of the U.S. ocean dumping policies dating back to 1972 in which a person might be interested. Probably one of the best single sources of information on the topic.

Cuéllar, Amanda D. and Michael E. Webber. 2010. "Wasted Food, Wasted Energy: The Embedded Energy in Food Waste in the United States." *Environmental Science and Technology* 44(16): 6464–6469. https://pubs.acs.org/doi/full/10.1021/es100310d. Accessed on July 25, 2019.

A factor that can easily be ignored about food waste is another type of waste: the loss of energy used in producing wasted foods. These authors estimate that about 2 percent of the nation's energy expenditure is lost as a result of food waste.

Esmaeilian, Behzad, et al. 2018. "The Future of Waste Management in Smart and Sustainable Cities: A Review and Concept Paper." *Waste Management* 81: 177–195.

The authors examine some of the deficiencies and other problems existing in today's waste management systems and then compare those problems with predictions for future waste management systems developed for sustainable cities.

Farady, Susan E. 2019. "Microplastics as a New, Ubiquitous Pollutant: Strategies to Anticipate Management and Advise Seafood Consumers." *Marine Policy* 104: 103–107.

Research indicates that microplastics are now being found in aquatic organisms consumed by humans as foods. The author explores the health issues posed by this discovery and suggests some actions that can be taken to protect consumers from the deleterious effects of the practice.

Filho, Walter Leal, et al. 2019. "Plastic Debris on Pacific Islands: Ecological and Health Implications." *Science of the Total Environment* 670: 181–187.

Plastic pollution of the ocean has been the subject of a considerable amount of research. Less studied is the effect of this phenomenon on the beaches and land areas of islands. This study finds that such effects are significant and need to be addressed.

Frosch, Robert A., and Nicholas E. Gallopoulos. 1989. "Strategies for Manufacturing." *Scientific American* 261(3): 144–152. http://isfie.onefireplace.com/resources/Documents/Strategies_For_Manufacturing_Sci_American_1989.pdf. Accessed on July 24, 2019.

This paper is often regarded as one of the seminal works in the field of valorization of wastes, the conversion of waste materials to products of greater economic value.

Gaba, Jeffrey M. 2012. "Exporting Waste: Regulations of the Export of Hazardous Wastes from the United States." *William & Mary Environmental Law and Policy Review* 36(2): 405–490. https://scholarship.law.wm.edu/cgi/viewcontent.cgi?referer=https://www.google.com/&httpsredir=1&article=1542&context=wmelpr. Accessed on July 10, 2019.

U.S. policy on the exportation of wastes from this country to other parts of the world has a long and checkered history that may or may not have taken into consideration the best interests of the receiving nation for such trash. This article provides a detailed review of that policy.

Gao, Ya, et al. 2019. "Global Trends and Future Prospects of E-waste Research: A Bibliometric Analysis." *Environmental Science and Pollution Research*, 26(17): 17809–17820.

The authors conduct a bibliographical search to discover "the status quo, hot topics, and future prospects in the field of e-waste."

Hartmann, Chris. 2018. "Waste Picker Livelihoods and Inclusive Neoliberal Municipal Solid Waste Management Policies: The Case of the La Chureca Garbage Dump Site in Managua, Nicaragua." *Waste Management* 71: 565–577.

What happens when a large dump, used by nearby waste pickers for their economic survival, is converted to a sanitary landfill? The author explores this question by interviewing more than 400 individuals in this category. He finds that economic survival and equity are very much changed by the new facility.

Havlíček, Filip. 2015. "Waste Management in Hunter-Gatherer Communities." *Journal of Landscape Ecology* 8(2): 47–59.

https://content.sciendo.com/view/journals/jlecol/8/2/article-p47.xml. Accessed on June 15, 2019.

Havlíček, Filip, and Kuča Martin. 2017a. "Waste Management at the End of the Stone Age." *Journal of Landscape Ecology* 10(1): 44–57. https://content.sciendo.com/view/journals/jlecol/10/1/article-p44.xml. Accessed on June 15, 2019.

Havlíček, Filip, and Kuča Martin. 2017b. "Waste Management in Bronze Age Europe." *Journal of Landscape Ecology* 10(1): 35–43. https://www.degruyter.com/downloadpdf/j/jlecol.2017.10.issue-1/jlecol-2017-0008/jlecol-2017-0008.pdf. Accessed on June 15, 2019.

Havlíček, Filip, and Miroslav Morcinek. 2016. "Waste and Pollution in the Ancient Roman Empire." *Journal of Landscape Ecology* 9(3): 33–49.

Havlíček, Filip, Adela Pokorná, and Jakub Zálešák. 2017. "Waste Management and Attitudes Towards Cleanliness in Medieval Central Europe." *Journal of Landscape Ecology* 10(3): 266–287.

Waste disposal by humans is a practice that dates back to prehistoric times. In this series of articles, environmental researcher Filip Havlíček and his colleagues review waste disposal and management technologies in five early periods: hunter-gatherer societies, the Stone Age and Bronze Age, ancient Rome, and Medieval Europe.

Heeres, R. R., W. V. Vermeulen, and F. B. de Walle. 2004. "Eco-industrial Park Initiatives in the USA and the Netherlands: First Lessons." *Journal of Cleaner Production* 12(8–10): 985–995. https://www.sciencedirect.com/science/article/pii/S0959652604000873. Accessed on July 24, 2019.

The authors study and compare three eco-industrial parks in the Netherlands and the United States. They conclude that the former are somewhat more successful, largely because they have the aid and the backing of the national government rather than local or regional entities.

Hidalgo, D., J. M. Martín-Marroquín, and F. Corona. 2019. "A Multi-waste Management Concept as a Basis Towards a Circular Economy Model." *Renewable and Sustainable Energy Reviews* 111: 481–489.

The authors describe and discuss a "multi-waste plant," in which a variety of types of wastes are treated in the same facility. They describe how this type of arrangement is a powerful tool in the circular model of waste management.

Horton, Alice A., et al. 2017. "Microplastics in Freshwater and Terrestrial Environments: Evaluating the Current Understanding to Identify the Knowledge Gaps and Future Research Priorities." *Science of the Total Environment* 586: 127–141. https://www.researchgate.net/publication/313358945_Microplastics_in_freshwater_and_terrestrial_environments_Evaluating_the_current_understanding_to_identify_the_knowledge_gaps_and_future_research_priorities. Accessed on July 27, 2019.

Plastic and microplastic pollution is a relatively well-studied problem. Much less is known about the pollution of lakes, streams, and other freshwater bodies. This article summarizes the knowledge available so far on the topic and areas of research that should be considered for the future.

Jambeck, Jenna R., et al. 2015. "Plastic Waste Inputs from Land into the Ocean." *Science* 347(6223): 768–770. https://wedocs.unep.org/bitstream/handle/20.500.11822/17969/Plastic_waste_inputs_from_land_into_the_ocean.pdf. Accessed on July 26, 2019.

This research team attempts to estimate the amount of plastic wastes released into the oceans from land-based sources. They conclude that the problem of plastic wastes could increase by as much as ten times by 2025.

Kabongo, Jean D. 2010. Strategic Challenges of Creating Value with Wastes: What's Inside Residual Materials Reclamation?" *International Journal of Sustainable Strategic Management* 2(2): 184–203.

The author explores the use of valorization of wastes as practiced by twelve Canadian firms and draws some general conclusions from these observations.

Kivenson, Veronika, et al. 2019. "Ocean Dumping of Containerized DDT Waste Was a Sloppy Process." *Environmental Science & Technology* 53(6): 2971–2980.

At one time, ocean dumping was the primary method for getting rid of hazardous chemical wastes, such as the toxic pesticide DDT. The researchers here used a sophisticated method for locating DDT at various levels in the ocean and distinguished between ocean dumping and land run-off sources of the chemicals. They found the former far more responsible for current ocean pollution than the latter.

Kore, D. Sudarshan. 2019. "Sustainable Utilization of Plastic Waste in Concrete Mixes—a Review." *Journal of Building Materials and Structures* 5(2): 212–217. https://www.researchgate.net/publication/330555546_Sustainable_Utilization_of_Plastic_Waste_in_Concrete_Mixes-a_Review. Accessed on July 8, 2019.

The conversion of plastic wastes to useful products has been a particularly difficult challenge for researchers. This article summarizes some of the suggestions that have been made for using waste plastics in the formation of concrete materials.

Lebreton, L., et al. 2018. "Evidence That the Great Pacific Garbage Patch Is Rapidly Accumulating Plastic." *Scientific Reports* 8(1): 1–15. https://www.nature.com/articles/s41598-018-22939-w.pdf. Accessed on July 27, 2019.

This report focuses on studies of the source, amount, and character of plastic waste pollution in the Pacific Ocean. It deals primarily with the region generally known as the Great Pacific Garbage Patch. The research team reports that the amount of debris in the region is four to sixteen times as great as previously reported.

Lebreton, Laurent, and Anthony Andrady. 2019. "Future Scenarios of Global Plastic Waste Generation and Disposal." *Palgrave Communications* 5(1): 1–11. https://www.researchgate.net/publication/330542951_Future_scenarios_of_global_plastic_waste_generation_and_disposal. Accessed on July 26, 2019.

Using high-resolution observational technology, the authors project the status of plastic pollution in the oceans in 2060. They conclude that developing nations are at the greatest risk of harmful social and economic effects by their projections. They suggest some ways by which worst-case scenarios can be avoided in such regions.

Levis, J. W., et al. 2010. "Assessment of the State of Food Waste Treatment in the United States and Canada." *Waste Management* 30(8–9): 1486–1494. http://www.seas.columbia.edu/earth/wtert/sofos/Food_Waste_Web_Barlaz_Themelis.pdf.

An estimated 97 percent of all food waste is deposited in landfills. Researchers are not exploring alternative methods of using wastes to produce energy and useful products. This article reports on a survey of studies being conducted in the United States and Canada on this topic.

Lindenlauf, Astrid. 2003. "The Sea as a Place of No Return in Ancient Greece." *World Archaeology* 35(3): 416–433.

For the ancient Greeks, dumping of wastes into the oceans, lakes, and rivers was a common practice because people thought these waterways were places of "no return" and as such was safe as a place of disposal. The author analyzes this view of water bodies in their relationship to waste disposal.

Louis, Garrick. 2004. "A Historical Context of Municipal Solid Waste Management in the United States." *Waste Management & Research* 2(4): 306–322. https://www.researchgate.net/publication/8253891_A_Historical_Context_of_Municipal_Solid_Waste_Management_in_the_United_States. Accessed on July 27, 2019.

This article traces important events in the development of waste management practices in the United States from the mid-19th century to the present day.

Milnes, Iain. 2017. "Why Are Corporations and Municipalities Moving to Zero Waste?" *Occupational Health & Safety* 86(4): EP12–EP13. http://www5.ohsonline.com/Articles/2017/02/22/Why-Are-Corporations-and-Municipalities-Moving-to-Zero-Waste.aspx. Accessed on July 29, 2019.

The author reviews the reasons that food companies are feeling the pressure to move toward zero waste programs and some methods that have been developed to achieve that objective.

Mohapatra, Sonali, et al. 2019. "Engineering Grass Biomass for Sustainable and Enhanced Bioethanol Production." *Planta* 250(2): 395–412.

Several ideas have been proposed for the use of agriculture waste or agricultural products in the synthesis of ethanol. This article discusses in some detail one such possible method.

Moharir, Rucha V., and Sunil Kumar. 2019. "Challenges Associated with Plastic Waste Disposal and Allied Microbial Routes for Its Effective Degradation: A Comprehensive Review." *Journal of Cleaner Production* 208: 65–76.

The authors review the status of the world's plastic waste problems and the effectiveness of existing technologies to deal with this problem. They conclude that other approaches are needed and review the methodology and examples of microbial degradation for dealing with the problem.

Nhat, P. Vo Hoang, et al. 2018. "Can Algae-Based Technologies Be an Affordable Green Process for Biofuel Production and Wastewater Remediation?" *Bioresource Technology* 256: 491–501.

The use of various types of algae for the remediation of wastes has been suggested. This article explores the technical issues involved in such a procedure and assesses the possibility of using such a technology for the production of biofuels from wastes.

Obimakinde, Samuel, et al. 2017. "Veterinary Pharmaceuticals in Aqueous Systems and Associated Effects: An Update." *Environmental Science and Pollution Research* 24(4): 3274–3297.

Animal wastes from farms, dairies, and other agricultural facilities often contain pharmaceutical products that may enter the nearby environment. This article reviews possible health issues associated with this process and methods for reducing its impact.

Patil, Pooja M., et al. 2019. "Conversion of Organic Biomedical Waste into Value Added Product Using Green Approach." *Environmental Science and Pollution Research* 26(7): 6696–6705.

Disposal of medical wastes is an especially challenging problem in waste management. This article describes a mechanism by which such wastes can be converted to fertilizers that can be used in a cost-effective way in the fields.

Perkins, Davin N., et al. 2014. "E-waste: A Global Hazard." *Annals of Global Health* 80(4): 286–295. https://www.sciencedirect.com/science/article/pii/S2214999614003208.

Concerns are being expressed by the possible threats posed to the environment and human health by e-wastes. This article describes a research study attempting to quantify the nature of that threat worldwide.

Scoble, Malcolm, Bern Klein, and W. Scott Dunbar. 2003. "Mining Waste: Transforming Mining Systems for Waste Management." *International Journal of Surface Mining, Reclamation and Environment* 17(2): 123–135. https://www.researchgate.net/publication/248913921_Mining_Waste_Transforming_Mining_Systems_for_Waste_Management. Accessed on July 24, 2019.

The authors argue that mining systems need to be redesigned based on the principle that "mining is a business whose success is fundamentally dependent upon waste management."

Sharma, Bhavisha, et al. 2019. "Recycling of Organic Wastes in Agriculture: An Environmental Perspective." *International Journal of Environmental Research* 13(2): 409–429.

Agricultural wastes are a rich source of compounds that can be used in farming and other agricultural applications. The authors consider mechanisms by which those compounds can be extracted from those wastes in environmentally sensitive ways.

Taghizadeh-Alisaraei, Ahmad, et al. 2017. "Potential of Biofuel Production from Pistachio Waste in Iran." *Renewable and Sustainable Energy Reviews* 72: 510–522. https://www.researchgate.net/publication/312525264_Potential_of_biofuel_production_from_pistachio_waste_in_Iran. Accessed on July 9, 2019.

The term *waste* can sometimes refer to the single product resulting from some specific manufacturing process rather than a collection of materials disposed of after use by humans. This article explores the former of these situations in which the question raised is how the wastes from a single operation, pistachio nut production, can be profitably disposed of.

Venkat, Kumar. 2011. "The Climate Change and Economic Impacts of Food Waste in the United States." *International Journal of Food System Dynamics* 2(4): 431–446. http://centmapress.ilb.uni-bonn.de/ojs/index.php/fsd/article/view/247/182. Accessed on July 25, 2019.

The author attempts to measure the economic impacts, as well as the effect on climate change, of food waste in the United States. He concludes that food waste accounts for about 29 percent of all food production and that the disposal of food wastes accounts for about 2 percent of the nation's annual greenhouse gas production.

Wang, Wanli, et al. 2019. "Current Influence of China's Ban on Plastic Waste Imports." *Waste Disposal & Sustainable Energy* 1(1):

67–78. https://link.springer.com/article/10.1007/s42768-019-00005-z. Accessed on July 24, 2019.

For many years, the United States, European Union, and other developed countries have relied on China's acceptance and recycling of plastic wastes for dealing with a major waste management problem. The authors here reviews actions being taken and planned by those nations to adjust to China's recent decision to withdraw from the plastic recycling market.

Yannopoulos, Stavros, et al. 2017. "History of Sanitation and Hygiene Technologies in the Hellenic World." *Journal of Water Sanitation and Hygiene for Development* 7(2): 163–180. https://www.researchgate.net/publication/313731172_History_of_sanitation_and_hygiene_technologies_in_the_Hellenic_World. Accessed on July 28, 2019.

This well-illustrated article provides a detailed look at techniques for sanitation systems in the world of ancient Greece.

Reports

Bogner, Jean, et al. 2007. "Waste Management." *Climate Change 2007: Mitigation of Climate Change*. Intergovernmental Panel on Climate Change, chapter 10. https://www.ipcc.ch/site/assets/uploads/2018/02/ar4-wg3-chapter10-1.pdf. Accessed on July 7, 2019.

Waste disposal is now generally recognized as a major contributor to climate change. This chapter reviews evidence for that position and many suggestions for dealing with this problem worldwide.

Donahue, Marie. 2018. "Waste Incineration: A Dirty Secret in How States Define Renewable Energy." Institute for Local Self-Reliance. https://ilsr.org/wp-content/uploads/2018/12/ILSRIncinerationFInalDraft-6.pdf. Accessed on July 25, 2019.

The author of this report draws three major conclusions from the legal status of waste incineration in states across the country: (1) the economics of waste incineration plants don't add up; (2) incinerators provide a classic case of environmental injustice; and (3) "renewable" trash burning is a legal oxymoron.

"Environmental Justice Progress Report." Annual report. 2017. Environmental Protection Agency. https://www.epa.gov/environmentaljustice/annual-environmental-justice-progress-reports. Accessed on July 24, 2019.

The EPA issues an annual report on the progress that has been made in implementing environmental justice principles into the agency's activities.

"Food: Material-Specific Data." 2019. Environmental Protection Agency. https://www.epa.gov/facts-and-figures-about-materials-waste-and-recycling/food-material-specific-data#FoodTableandGraph. Accessed on July 25, 2019.

This report provides the most recent summary of data on food generation and disposal methods for 1960 through 2015. About 76 percent of all food waste produced ends up in a landfill, 5 percent is composted, and 19 percent is incinerated.

Kaza, Silpa, et al. 2018. "What a Waste 2.0: A Global Snapshot of Solid Waste Management to 2050." Washington, DC: World Bank. https://openknowledge.worldbank.org/handle/10986/30317. Accessed on July 8, 2019.

This report provides a detailed overview of the current status of solid waste management in countries around the world. It projects changes likely to occur by 2030 and by 2050. It covers topics such as waste management costs, revenues, and tariffs; special wastes; regulations; public communication; administrative and operational models; and the informal sector.

Locock, Katherine E. S., et al. *The Recycled Plastics Market: Global Analysis and Trends*. Australia: CSIRO: Commonwealth Scientific and Industrial Research Organization.

This study attempts to outline the general features of the world's oceans' plastic pollution problem with special emphasis on the economic factors involved in finding ways of dealing with the problem.

"Medical Waste Management." 2011. International Committee of the Red Cross. https://www.icrc.org/en/doc/assets/files/publications/icrc-002-4032.pdf. Accessed on July 24, 2019.

This report provides a general review of the problems related to medical wastes and standard procedures for dealing with collection, storage, transport, treatment, and disposal.

"Medical Waste Management Market Overview." 2019. Prescient & Strategic Intelligence. https://www.psmarketresearch.com/market-analysis/medical-waste-management-market. Accessed on July 8, 2019.

The medical waste market is in flux today largely because of advances in medical technology, which have resulted in increasing quantities of waste materials and governmental regulations to guarantee control over the fate of these wastes. This report provides details about the status of the medical waste market in a number of categories on a worldwide basis.

"Marine Plastic Debris and Microplastics: Global Lessons and Research to Inspire Action and Guide Policy Change." 2016. Nairobi: United Nations Environment Programme. https://wedocs.unep.org/bitstream/handle/20.500.11822/7720/-Marine_plasctic_debris_and_microplastics_Global_lessons_and_research_to_inspire_action_and_guide_policy_change-2016Marine_Plastic_Debris_and_Micropla.pdf. Accessed on July 27, 2019.

In 2014, the United Nations Environment Assembly recommended a study of worldwide issues associated with the presence of plastics and microplastics in the world's oceans. This report summarizes the findings of the committee appointed to conduct that study, along with a group of more than fifty recommendations for actions dealing with the problem.

"Microplastics in the Great Lakes: Workshop Report." 2016. International Joint Commission. https://legacyfiles.ijc.org/tinymce/uploaded/Microplastics_in_the_Great_Lakes_Workshop_Report_FINAL_September14-2016.pdf. Accessed on July 27, 2019.

This document is the report of a binational (Canada and the United States) commission created to study issues of microplastics in the Great Lakes. The commission came up with a set of ten recommendations for better understanding and dealing with the problem of microplastic pollution in the Great Lakes.

"National Overview: Facts and Figures on Materials, Wastes and Recycling." 2018. Environmental Protection Agency. https://www.epa.gov/facts-and-figures-about-materials-waste-and-recycling/national-overview-facts-and-figures-materials. Accessed on July 24, 2019.

The EPA annually provides basic data and statistics on all aspects of waste management. This page is the most recent report available as of late 2020.

"Plan EJ 2014." 2011. Environmental Protection Agency. https://nepis.epa.gov/Exe/ZyPDF.cgi/P100DFCQ.PDF. Accessed on July 24, 2019.

In 2011, the Environmental Protection Agency announced a project designed to find and implement ways of incorporating the principles and practice of environmental justice into all of the department's activities. This website provides the details of the effort. A summary of the program's accomplishments can be found at https://www.federalregister.gov/documents/2013/05/09/2013-10945/epa-activities-to-promote-environmental-justice-in-the-permit-application-process#h-13. Accessed on July 24, 2019.

"Toxicological Profiles." 2019. Agency for Toxic Substances and Disease Registry. https://www.atsdr.cdc.gov/toxprofiledocs/index.html. Accessed on July 24, 2019.

ATSDR is the nation's premier source on the health effects of toxic substances. This web page provides an index to all the substances listed in its database. The report is an invaluable source of information about all substances generally found in hazardous wastes.

"Waste Atlas Report." Annual. D-Waste. Various links, as noted below.

The first World Atlas Report appeared in 2013 (https://www.iswa.org/fileadmin/galleries/News/WASTE_ATLAS_2013_REPORT.pdf, accessed on July 7, 2019). It was designed to make available "user-friendly, well-structured . . . to those who need it?" A simple way to do that, organizers decided, would be to organize existing data in a map of the world outlining data for each country. The report was issued again in 2014 (https://www.resource-recovery.net/sites/default/files/waste-atlas-report-2014-webedition.pdf, accessed on July 7, 2019) and then annually in map form ever since. The annual report (http://www.atlas.d-waste.com/, accessed on July 7, 2019) is almost certainly the most complete collection of waste management available anywhere today.

Internet

"Administrative Procedures Act Rules." 2019. Mississippi Department of Environmental Quality. http://www.sos.ms.gov/ACCode/00000517c.pdf. Accessed on July 9, 2019.

States, counties, cities, towns, and other administrate entities have a host of laws and regulations dealing with all aspects of solid waste management. This page is an example of one way in which one state deals with one specific aspect of waste management, the transportation of such materials within the state. Similar regulations can be found throughout internet sources.

Agamuthu, P., et al. 2019. "Marine Debris: A Review of Impacts and Global Initiatives." *Waste Management & Research*. https://doi.org/10.1177/0734242X19845041. https://journals.sagepub.com/doi/abs/10.1177/0734242X19845041. Accessed on July 9, 2019.

Solid wastes have become a major environmental issue for the world's oceans and coastal environments. The authors discuss the details of this solid waste management problem and some major regional and international efforts that have been developed to deal with the problem.

"Alternative Uses for Tailings." 2019. Weir. https://www.global.weir/newsroom/news-articles/alternative-uses-for-tailings/. Accessed on July 24, 2019.

Tailings remain one of the most serious problems in mining waste management. This article discusses some methods of dealing with this problem.

"Animal Waste Management Software." 2019. U.S. Department of Agriculture. https://www.nrcs.usda.gov/wps/portal/nrcs/detailfull/national/landuse/crops/npm/. Accessed on July 7, 2019.

Livestock and poultry operations are a major source of agricultural waste products. These products pose a significant disposal problem for farmers, but they also have the potential for significant benefits to agriculture. This website provides access to a software program that can be used to predict the outcome in a number of areas relating to animal wastes. For a further discussion of this issue, also see "What is an Animal Waste Management Plan?" 2019. Rutgers. New Jersey Agricultural Station. https://njaes.rutgers.edu/animal-waste-management/what-is-an-awmp.php. Accessed on July 7, 2019.

Aznar-Sánchez, José A., et al. 2018. "Mining Waste and Its Sustainable Management: Advances in Worldwide Research." *Minerals* 8(7): 284; https://doi.org/10.3390/min8070284. https://www.mdpi.com/2075-163X/8/7/284. Accessed on July 24, 2019.

This study examines research conducted on mining wastes and sustainable development worldwide between 1988 and 2017.

Baker, Maverick. 2018. "How to Eliminate Plastic Waste and Plastic Pollution with Science and Engineering." Interesting Engineering. https://interestingengineering.com/how-to-eliminate-plastic-waste-and-plastic-pollution-with-science-and-engineering. Accessed on July 28, 2019.

Finding ways to convert waste plastics into useful new materials and objects is a major goal of much research today. This web page discusses about a dozen of the most promising ideas that have been developed thus far.

Baptista, Ana. 2018. "Garbage In, Garbage Out: Incinerating Trash Is Not an Effective Way to Protect the Climate or Reduce Waste." Conversation. https://theconversation.com/garbage-in-garbage-out-incinerating-trash-is-not-an-effective-way-to-protect-the-climate-or-reduce-waste-84182. Accessed on July 25, 2019.

The author explains the fundamentals of waste incineration and then explains why that process is not a wise choice for waste disposal.

"Basic Concepts of Integrated Solid Waste Management." 2012. International Best Practices Guide for LFGE Projects. https://www.globalmethane.org/documents/toolsres_lfg_IBPGch1.pdf. Accessed on July 6, 2019.

This review sets out the basic principles that underlie integrated solid waste management systems, with special focus on the release of methane gas as a contributor to global warming.

"Basic Information about Landfill Gas." 2019. Environmental Protection Agency. Accessed on July 6, 2019.

A topic that sometimes receives an inadequate amount of attention in waste management discussions is the gases that are produced as a result of waste decomposition in landfills, the effects of the release of those gases on the environment, and the ways in which the problem can be solved.

Bishop, Steven. 2019. "Hospital Infection Control: Waste Management." Infectious Disease Advisor. https://www.infectiousdiseaseadvisor.com/home/decision-support-in-medicine/hospital-infection-control/waste-management/. Accessed on July 7, 2019.

Health-care facilities generate a substantial amount of waste materials, many of which carry with them the agents that cause infectious diseases. This article examines in detail what those diseases are, how they can be transmitted through waste products, and what can be done to reduce the risk of infection from wastes.

"Cell Phone Toxins and the Harmful Effects on the Human Body When Recycled Improperly." 2013. e-Cycle. https://www.e-cycle.com/tag/e-waste-effects-to-the-human-body/. Accessed on July 24, 2019.

Experts often warn about the potential health risks posed by disposal of electronic wastes. This brief article describes some of the toxins involved in such a problem.

Chew, Kit Wayne, et al. 2019. "Transformation of Biomass Waste into Sustainable Organic Fertilizers." *Sustainability* 11(8):

2266. https://doi.org/10.3390/su11082266. https://www.researchgate.net/publication/332434018_Transformation_of_Biomass_Waste_into_Sustainable_Organic_Fertilizers. Accessed on July 8, 2019.

One long-term element of waste management has involved the conversion of organic matter in wastes to fertilizers. This article reviews the current state of affairs in this field of research.

Chitnavis, Samir. 2019. "Fueling the Future? The Rise and Fall of Biofuels." Science Entrepreneur Club. https://www.science-entrepreneur.com/blog-1/fueling-the-future. Accessed on July 24, 2019.

Biofuels have been touted as a critical element in dealing with the world's waste management problems. This article provides a review of the history of biofuel development and its possible applications in the future.

"Cholera in London." 2019. Cholera and the Thames. https://www.choleraandthethames.co.uk/cholera-in-london/origins-of-cholera/. Accessed on July 8, 2019.

This collection of articles tracks the series of cholera epidemics that struck London in the mid- to late 19th century and the impact those events had on the development of a public health consciousness in the city.

Cioca, Bagriela, and Florentina-Daniela Munteanu. 2019. "Estimation of the Amount of Disposed Antibiotics." *Sustainability* 11(6): 1800. https://doi.org/10.3390/su11061800. https://www.researchgate.net/publication/332017617_Estimation_of_the_Amount_of_Disposed_Antibiotics. Accessed on July 9, 2019.

The authors discuss the problems of estimating the amount of antibiotics that may occur in municipal solid wastes and the reasons these data are of importance in waste management and public health systems.

"Compost Fundamentals." n.d. Washington State University. http://whatcom.wsu.edu/ag/compost/fundamentals/index.htm. Accessed on July 27, 2019.

One of the best resources available on all aspects of composting in the process of waste management. The website covers every topic from aerobic and anaerobic decomposition to carbon-nitrogen relationships to fly control to economic aspects.

Danigelis, Alyssa. 2019. "TerraCycle Launches Loop Circular Delivery Service with Major Brands." Environmental Leader. https://www.environmentalleader.com/2019/01/terracycle-circular-delivery-loop/. Accessed on July 29, 2019.

The recycling organization, TerraCycle, has introduced a new method of shipping products that makes use of recyclable containers rather than cardboard boxes. Several major companies have already signed on to participate in the program.

"Energy Recovery from the Combustion of Municipal Solid Waste (MSW)." 2019. Environmental Protection Agency. https://www.epa.gov/smm/energy-recovery-combustion-municipal-solid-waste-msw. Accessed on July 25, 2019.

This website provides a good overall introduction to waste incineration as a source of energy, combustion technologies, the history of waste-to-energy, and frequent questions about the procedure.

"Food Loss and Waste in the United States and Worldwide." 2016. Hunger Notes. https://www.worldhunger.org/food-loss-and-waste-in-the-united-states-and-worldwide/. Accessed on July 25, 2019.

The terms *food loss*" and *food waste* do not mean the same thing. This article explains the difference between the two as well as the data for each and the ways in which each event occurs.

Fotedar, Amita. 2017. "Composting—a Solution to the Burning Problem of Solid Waste Management." https://permaculturenews.org/2017/06/23/composting-solution-burning-problem-solid-waste-management/. Accessed on July 27, 2019.

This web page provides a good general introduction to compositing as a method of waste management with helpful links to related websites.

"Franklin Eco-Industrial Park." 2019. http://franklinecopark.com/. Accessed on July 24, 2019. The Franklin Eco-industrial Park is one of the first of its kind in the United States. This web page provides a general introduction and overview of the project.

Fuller, Thomas. 2019. "In San Francisco, Making a Living from Your Billionaire Neighbor's Trash." *New York Times*. https://www.nytimes.com/2019/04/07/us/trash-pickers-san-francisco-zuckerberg.html. Accessed on July 29, 2019.

Trash-picking as a way of making a living is generally thought of as a Third World occupation. Such ways of life now exist in the United States also, to a large extent because of the huge gap in income between the richest and poorest members of society. This article discusses trash-picking by individuals who live within a few miles of some of the richest individuals on the planet.

Geyer, Roland, and Jenna Jambeck. 2017. "Production, Use, and Fate of All Plastics Ever Made." *Science Advances* 3(7): e1700782. https://doi.org/10.1126/sciadv.1700782. https://www.researchgate.net/publication/318567844_Production_use_and_fate_of_all_plastics_ever_made. Accessed on July 28, 2019.

The authors attempt to estimate the total amount of plastic ever made and the fate of that plastic. They estimate that 8,300 million metric tons of plastic were made prior to 2017, of which 9 percent was recycled, 12 percent was incinerated, and 79 percent left in landfills or the natural environment.

"A Guide to Plastic in the Ocean." 2019. National Ocean Service. https://oceanservice.noaa.gov/hazards/marinedebris/plastics-in-the-ocean.html. Accessed on July 26, 2019.

This relatively short introduction to plastic pollution of the oceans is especially useful because of the many links it provides to other articles and internet sources dealing with specific topics on the subject.

"History of Solid Waste Management." 2013. Environmentalists Every Day. https://archive.is/20131024012358/http://www.environmentalistseveryday.org/publications-solid-waste-industry-research/information/history-of-solid-waste-management/early-america-industrial-revolution.php. Accessed on July 27, 2019.

This website provides an interesting and detailed account of the history of waste management in the United States from 1657 to 1899.

"How Did Nomadic Tribes in North America Pre-1850 Deal with Human Waste Disposal at Large Summer Encampments?" 2014. Quora. https://www.quora.com/How-did-nomadic-tribes-in-North-America-pre-1850-deal-with-human-waste-disposal-at-large-summer-encampments. Accessed on July 28, 2019.

Very little is known about waste management practices among pre-Columbian Native American tribes. This short web page provides some insights into some possible information on these practices.

Huang, Ximin (Natalie), Atalay Atasu, and Beril L. Toktay. 2015. "Design Implications of Extended Producer Responsibility for Durable Products." Georgia Tech Scheller College of Business Research Paper No. 2015–17. https://doi.org/10.2139/ssrn.2693152. https://papers.ssrn.com/sol3/papers.cfm. Accessed on July 8, 2019.

This paper reviews one of the options available for implementation of the waste management system known as extended producer responsibility. See more on this topic under *Extended Producer Responsibility: A Guidance Manual for Governments* (2001), under "Books" above.

"Indonesia Returns Millions of Tonnes of Waste to Australia." 2019. YouTube. https://www.youtube.com/watch?v=ySpR0jrDk5c. Accessed on July 11, 2019.

This video provides a nice update on the practice of developed nations' sending their trash to developing nations. Among the interesting features is the fact that such trash often contains items that were supposedly not covered by a deal between the two countries, such as the presence of dirty diapers in a shipment of (supposedly only) waste paper.

Joyce, Christopher. 2019. "Where Will Your Plastic Trash Go Now That China Doesn't Want It?" NPR. https://www.npr.org/sections/goatsandsoda/2019/03/13/702501726/where-will-your-plastic-trash-go-now-that-china-doesnt-want-it. Accessed on July 24, 2019.

This article discusses the history of China's involvement with the world's solid waste disposal and the problems created by its decision to not accept these wastes any longer.

Kilcarr, Sean. 2012. "Far and Away: A Look at Long-Haul Waste Transport." Waste 360. https://www.waste360.com/long-haul/far-and-away-look-long-haul-waste-transport. Accessed on July 9, 2019.

Over the past few decades, regulatory changes and other factors have led to the siting of solid waste landfills at greater distances from urban areas where those wastes are produced. This article reviews some of the characteristics of today's solid waste transportation problems and solutions.

"Learn about Ocean Dumping." 2018. Environmental Protection Agency. https://www.epa.gov/ocean-dumping/learn-about-ocean-dumping. Accessed on July 27, 2019.

This comprehensive website covers most aspects of ocean dumping, including traditional use of the practice for getting rid of wastes; the Marine Protection, Research, and Sanctuaries Act of 1972 and its current implementation; and current regulations dealing with ocean dumping of wastes.

LeBlanc, Rick. 2017. "E-waste State of the Union." Investment Recovery Association. https://invrecovery.org/e-waste-state-of-the-union-recycling-facts-figures-and-the-future/. Accessed on July 24, 2019.

This article provides several facts and statistics about the status of e-wastes in the United States as of 2019.

LeBlanc, Rick. 2018. "What Is a Waste Transfer Station?" Small Business. https://www.thebalancesmb.com/what-is-a-waste-transfer-station-2877735. Accessed on July 9, 2019.

One of the characteristic features of a modern solid waste management system is a storage facility at which wastes can be kept in the chain of events from waste collection to final resting stage (e.g., landfill or incineration site). This article discusses one such storage facility, a transfer station.

Newman, Sarach. 2015. "Rethinking Refuse: A History of Maya Trash." Academic.edu. https://www.academia.edu/15299412/Rethinking_Refuse_A_History_of_Maya_Trash. Accessed on July 6, 2019. (Be patient. This is a large file.)

For her doctoral dissertation, the author does a very detailed and extensive review of the role of waste management in the Mayan civilization.

"Ocean Dumping Management Timeline." 2018. Environmental Protection Agency. https://www.epa.gov/ocean-dumping/ocean-dumping-management-timeline. Accessed on July 27, 2019.

This website lists, but does not discuss, major events in the U.S. history of ocean dumping actions. The list provides a good beginning for a more detailed study of various topics, however.

"Ocean Pollution." 2019. MarineBio. https://marinebio.org/conservation/ocean-dumping/. Accessed on July 27, 2019.

This lengthy web page contains sections dealing with virtually every aspect of ocean pollution, including mercury contamination, runoff, and pollution; waste in the ocean; sewage sludge; radioactive wastes; ocean dumping; and the ocean ecosystem.

Pietzsch, Natália, José Luis Duarte Ribeiro, and Janine Fleith de Medeiros. 2017. "Benefits, Challenges and Critical Factors of Success for Zero Waste: A Systematic Literature Review." *Waste Management* 67: 324–353.

As the title indicates, this review of the literature attempts to identify major benefits and challenges to adopting zero waste practices, along with factors that may contribute to or detract from the success of such endeavors, such as consumer attitudes and behaviors and economic factors.

"Regulated Medical Wastes—Overview." 2015. Pollution Prevention and Compliance Assistance Information for the Healthcare Industry. http://www.hercenter.org/rmw/rmwoverview.php. Accessed on July 24, 2019.

This website provides a good general overview of the nature of medical wastes and the ways in which they are regulated by state and federal law.

Rodríguez, José Remesal. 1997. "Exhibition: Mount of Amphorae." Wayback Machines. https://web.archive.org/web/20050302200446/http://ceipac.gh.ub.es/MOSTRA/u_expo.htm. Accessed on July 8, 2019.

This superb website brings together several papers written on the history of Mount Testaccio, in Rome, one of the most famous "garbage heaps" in all of the classical world.

The articles are accompanied by excellent drawings and photographs to illustrate the subjects of the papers.

Saleh, Hosam El-Din M. 2016. "Introductory Chapter: Introduction to Hazardous Waste Management, Management of Hazardous Wastes." In Hosam El-Din M. Saleh and Rehab O. Abdel Rahman, eds. IntechOpen. https://doi.org/10.5772/64245. https://www.intechopen.com/books/management-of-hazardous-wastes/introductory-chapter-introduction-to-hazardous-waste-management. Accessed on July 28, 2019.

This chapter provides a good general introduction to the topic of hazardous waste management with a discussion of the types of hazardous wastes and methods for treating such materials.

Sauer, Richard L., and George K. Jorgensen. 1975. "Waste Management System." National Aeronautics and Space Administration. https://history.nasa.gov/SP-368/s6ch2.htm. Accessed on July 7, 2019.

Disposal of urine and feces during space missions has always been a crucial, but seldom widely discussed topic. This article provides a detailed discussion of the earliest systems developed for dealing with this issue by U.S. astronauts.

Seavitt Nordenson, Catherine. 2016. "The Miasmist: George E. Waring, Jr. and the Evolution of Modern Public Health." Landscape Research Record No. 5. http://thecela.org/wp-content/uploads/SEAVITT-NORDENSON.pdf. Accessed on July 6, 2019.

This excellent article provides a detailed review of the life of Waring, focusing on his efforts to improve waste management systems in New York City in the late 19th and early 20th centuries.

"6 Negative Effects of Improper Waste Management." 2017. Metropolitan Transfer Station. https://www.metropolitantransferstation.com.au/blog/negative-effects-of-improper-waste-management. Accessed on July 9, 2019.

This web page focuses on six major ways in which solid wastes pollute the environment, including soil and water contamination, climate change, and harm toward animal and marine life.

Sofi, Massoud, et al. 2019. "Transforming Municipal Solid Waste into Construction Materials." *Sustainability* 11(9): 2661. https://doi.org/10.3390/su11092661. https://www.researchgate.net/publication/332995299_Transforming_Municipal_Solid_Waste_into_Construction_Materials. Accessed on July 9, 2019.

The authors take note of the harmful effects of solid wastes and their management methods on human health and the environment. They suggest that one way of reducing these effects is by converting solid wastes into construction materials by use of a process they explain in this article.

"Solid Waste Management." 2018. Island Institute. http://www.islandinstitute.org/what-works/solid-waste-management. Accessed on July 8, 2019.

Communities located on islands have some very specialized waste management issues. For example, there may not be room to set up a sanitary landfill or effective trash incinerator. This article describes the way in which one such community deals with these problems.

Solid Waste Management and Greenhouse Gases: A Life-Cycle Assessment of Emissions and Sinks. 3rd ed. 2006. Environmental Protection Agency. https://nepis.epa.gov/Exe/ZyPDF.cgi/60000AVO.PDF?Dockey=60000AVO.PDF. Accessed on July 6, 2019.

This document offers a detailed analysis of greenhouse gas emissions from landfills, their effect on the environment, and current strategies for reducing and eliminating such emissions.

"Solid Waste Management Plan." 2011. City of Seattle. https://www.seattle.gov/utilities/documents/plans/solid-waste-plans/solid-waste-mgmt-plan. Accessed on July 7, 2019.

Some municipalities have been more aggressive than others in terms of developing long-term programs for waste management. This detailed and extensive document adopted by the city of Seattle illustrates one such plan.

"South Carolina Solid Waste Policy and Management Act." 2019. https://www.scstatehouse.gov/code/t44c096.php. Accessed on July 9, 2019.

Every state has some type of legislation controlling the handling of solid wastes within the state. Most laws are extensive and detailed, covering all aspects of waste management, from generation to final disposal. This web page illustrates the law in one state, South Carolina, as an example. Similar laws in other states can also be found on the internet.

"Striving for Zero Waste." 2019. SF Environment. https://sfenvironment.org/striving-for-zero-waste. Accessed on July 29, 2019.

The city of San Francisco has adopted a challenging program designed to eliminate wastes from the city by 2020. This website describes the details of the program. Follow links to obtain more complete information about all aspects of the program.

"Toilets, Earth Closets, and House Plumbing." 2019. The History of Sanitary Sewers. http://www.sewerhistory.org/photosgraphics/toilets-earth-closets-and-house-plumbing/. Accessed on July 8, 2019.

This article provides a good general summary of the early history of the modern flush toilet.

"Tribal Waste Management Program." 2018. Environmental Protection Agency. https://www.epa.gov/tribal-lands/tribal-waste-management-program. Accessed on July 7, 2019.

Waste management issues on tribal lands may have unique problems and issues not found in nontribal areas. The EPA has an extensive program designed to help tribes identify those issues, learn about the resources available for dealing with those issues, and describing programs that can put those ideas into practice.

"Waste Management Infrastructure." 2019. Earthworks. https://earthworks.org/issues/waste_management_infrastructure/. Accessed on July 27, 2019.

The use of hydraulic fracturing ("fracking") as a way of collecting gas and oil from underground has become extremely popular in the past few decades. One of the major problems involved with this technology is what to do with the very substantial amounts of waste that are produced in fracking operations. This website reviews the nature of those problems and methods that have been developed to deal with them.

"Waste Valorization." n.d. Center for Energy Initiatives (CEI). https://www.aiche.org/cei/topics/energy/waste-valorization. Accessed on July 24, 2019.

This website includes many examples of waste valorization processes that occurred previously as webinars, conference presentations, or other events sponsored by CEI.

"What Is a Freegan?" 2019. Freegan.info. https://freegan.info/. Accessed on July 25, 2019.

Freeganism is a relatively recently developed way of life that is based on an approach to living "based on sharing resources, minimizing the detrimental impact of our consumption, and reducing and recovering waste and independence from the profit-driven economy." The very substantial value of waste food for human consumption is one important pillar of the group's philosophy and practices.

"Winning on Reducing Food Waste: FY 2019–2020 Federal Interagency Strategy." 2019. https://www.epa.gov/sites/production/files/2019-05/documents/reducingfoodwaste_strategy.pdf. Accessed on July 25, 2019.

In 2018, the U.S. Department of Agriculture, U.S. Environmental Protection Agency, and U.S. Food and Drug Administration joined together to create the Winning on Reducing Food Waste initiative, whose major goal it is to reduce food loss and waste by in the United States by 50 percent by 2030.

Introduction

Humans have been generating waste since time immemorial. Throughout that time, waste has often been simply "thrown away"—tossed onto the ground; burned and released into the atmosphere; or dumped into a river, lake, or other body of water. The disposal of raw wastes into the environment may have been acceptable at one or another time in history and place on Earth, but such has not always been the case. Indeed, with the development of human civilization, the quantity of waste and its effects on the world around us have become more problematic and profound. This chapter summarizes some of the most important events in the history of waste production, disposal, and treatment throughout history.

Prehistory Early humans use pits for the disposal of stone flakes and chips, animal bones, shells, ashes, pottery shards, slag from mining operations, and other waste materials. Determination as to the precise purpose of individual pits is difficult to determine at this point in time. But evidence seems clear in at least some cases that the pits were intended as a "final resting place" for materials and objects no longer intended for use.

ca. 3000 BCE The earliest known landfill is constructed in Knossos, capital of the ancient Minoan civilization. The landfill

Waste incinerators are often sited close to regions occupied by human settlement or agricultural operations. (Dutchscenery/Dreamstime.com)

consists of alternate layers of solid wastes and soil, much as is the case with modern landfills.

ca. 600 BCE Reference to a geographical location called Gehenna in the New Testament of the Bible describes a burning pit generally thought to have been a waste disposal site at which refuse was burned, although differences of opinion exist as to that interpretation.

ca. 500 BCE In response to a crisis of trash accumulation in the city, the government of Athens, Greece, adopts a law requiring that waste products be deposited in landfills to be constructed at least a mile from city limits.

800–900 CE A solid waste management method developed by the Mayans in the region of modern-day Guatemala involves the collection of all forms of solid wastes followed by their incineration in a large annual public ceremony.

1031 Japanese papermakers begin to use sheets of previously used paper for the production of new paper. This is thought to be perhaps the first example of paper recycling in history, although the tendency of the Japanese, in general, to adopt Chinese technology suggests that the latter may have used paper recycling at an even earlier date. No such evidence actually exists as of the present day, however.

1086 Early Chinese metallurgists use scrap iron for the cementation of copper, an early example of recycling. Cementation involves the precipitation of copper ions from a solution in the presence of elemental iron, thus recovering more of the product (copper) than would otherwise be possible. The process is still used today.

1185 King Philip Augustus of France orders that the streets of Paris be paved. He takes the action because residents of the city usually dispose of their household wastes by throwing them out the window. The king also issues an edict specifically prohibiting this practice. The French law is one of many acts in other countries designed to reduce the problem of urban solid wastes. Similar laws were issued in Florence in 1237, Sienna in 1290, and Bruges in 1332.

1220 The government of Naples adopts a law that declares that "who deposits muck or debris at other than the designated places is to be seized and sent on a galley or be whipped across the whole city."

1297 The British parliament passes a law requiring all homeowners to keep the area in front of their homes free of refuse. There is little evidence that families obeyed this law but, instead, continued to burn their wastes in open fires on the property.

1354 One of the world's first waste management systems is created in England when King Edward III creates the job of the "raker." The raker's job was to go through the streets, alleys, and byways of cities to collect refuse that had collected in their gutters. The job was particularly onerous (but necessary) because most households dumped their human as well as household wastes in gutters. Rakers are sometimes referred to as the world's first garbage collectors.

1388 The English Parliament passes what is thought to be one of the world's first laws on waste management, the Nuisances in Towns Act, forbidding the disposal of solid wastes in public ditches and waterways.

1400 Piles of trash begin to accumulate in most major cities, causing problems other than odor, unsightliness, and health problems. In Paris, for example, the accumulation of garbage outside city walls becomes so great that the military is unable to protect the city adequately because they can't fire their weapons over the tops of the trash heaps. In addition, enemy soldiers are able to climb on the trash pile to get over the walls and into the city.

1407 The English parliament adopts another piece of legislation in its outgoing effort to control the country's waste management problem. Homeowners are instructed to store their wastes at an indoor site until collected by crews of rakers. After collection, the wastes were to be transported to a central location where they were to be sold as compost for agricultural operations or dumped into marshes and other waterways.

15th century Cities in areas that constitute modern-day France, Germany, and other countries establish requirements that wagons bringing goods (such as agricultural products) into a city, upon their return trip, remove wastes from the city into the countryside.

1500s Cementation of copper with iron is introduced in the Rio Tinto region of Spain and is widely used there and in other regions to the present day.

1560 The so-called First Cleanliness Decree issued in Hamburg requires that the town's market squares be cleaned four times a year at public expense.

1588 In one of the earliest instances of the privilege accorded the sovereign, Queen Elizabeth I grants a patent to Sir John Spielman for "the sole gathering of all rags and other articles for making paper" for a period of ten years. The patent provides Spielman with a monopoly for an early commercial recycling process.

Early 17th century Early colonists in America make use of pigs to solve at least part of their waste management problems. Experts calculate that seventy-five pigs should be able to eat one ton of refuse per day. Of course, the pigs then leave behind other types of wastes with their own problems.

1775 Scottish watchmaker and inventor Alexander Cumming obtains a patent for a type of flush toilet with an S-shaped pipe below the toilet seat. The design forms the basis of all modern toilet systems in use today.

1826 American inventor Samuel Morey explores the possibility of using ethanol (ethyl alcohol) as fuel for an internal combustion engine that he invented. His research is largely ignored by his contemporaries and successors in automotive development.

1842 British attorney Edwin Chadwick publishes a "General Report on the Sanitary Conditions of the Labouring Population of Great Britain." The report explains that the spread of disease among the working class in Great Britain was caused not by miasmic conditions in the country, as was generally believed, but by poor sanitation. Chadwick recommended that homes and streets be cleared of refuse, drainage systems be improved, and a clean water system be developed. He also suggested appointing a public health officer in every city and town. Chadwick's report is sometimes described as "the beginning of the age of sanitation" in Europe.

1848 The British parliament passes the Public Health Act, a piece of legislation incorporating most of Edwin Chadwick's recommendations in his "General Report on the Sanitary Conditions of the Labouring Population of Great Britain."

1848 Parliament adopts the Metropolitan Sewers Act, which forbids the construction of new houses lacking indoor toilets and connections to municipal sewer systems.

1850 A special commission in the state of Massachusetts, under the direction of Boston historian and bookseller Lemuel Shattuck, releases its "Report of the Sanitary Commission of Massachusetts," which points out the poor waste management systems then in use in the city of Boston and their effects on human health there.

1854 English physician John Snow shows that an epidemic of cholera results from polluted drinking water used by residents of the affected neighborhood and that simply removing the handle from the pump through which that water is collected brings an end to the epidemic.

1855 The approximate date on which the third, or so-called Modern Plague, began in China.

1858 The combination of an unusually warm summer and a poor sanitation system results in an event that became known as the Great Stink.

1874 The world's first trash incinerator is built in Nottingham, England.

1875 England's Public Health Act of 1875 updates the 1848 version of that legislation and changes the focus of public health activities from the nation as a whole to individual properties.

1882 The U.S. Congress passes the Rivers and Harbors Act, making it illegal to dump any solid waste into any navigable waterway in the country.

1896 Henry Ford's "Quadricycle" motor vehicle is designed to run on ethanol, gasoline, or some combination of the fuels. Gasoline wins out as the most popular fuel because of its lower cost.

1897 The city of New York creates what is probably the world's first municipal solid waste recycling center, the so-called picking yards. Wastes deposited there are sorted by value for reuse and provided to those who can make use of them in those ways.

1912 The first sanitary landfill is constructed in England, where the method is better known as *controlled tipping*. The system depends on the natural process of anaerobic decomposition for destruction of wastes. (Differences of opinion exist as to the date of the first such system, with others claiming 1915, 1916, 1920, or some other date.)

1937 The first sanitary landfill in the United States is opened in Fresno, California.

1944 Researchers at the Dow Chemical Company invent a new form of long-lasting, sturdy plastic, which they call Styrofoam.

1948 The Federal Water Pollution Control Act of 1948 provides the first modern legislation limiting the deposit of solid wastes into the nation's lakes and rivers.

1948 The city of New York opens a new landfill along the banks of the Fresh Kills estuary in Staten Island. Originally planned as a temporary depository, the site grew by 1955 to become the largest landfill in the world, covering more than 2,200 acres. It retained that title until it was closed in 2001, with plans to convert the buried waste depository into a public park (which was eventually achieved).

1953 The Niagara Falls (New York) School Board purchases a seventy-acre piece of property known as the Love Canal property for the construction of a new school. The property had formerly been used by the Hooker Chemical Company as a dump for its hazardous wastes. Soon after the sale, people began complaining about adverse health effects, which were eventually attributed to toxic chemicals buried at the site. The location is eventually declared a Superfund site, residents are paid to move out of the area, and the cleanup operation is completed in 2004.

1954 The city of Olympia, Washington, becomes the first municipality in the United States to pay consumers for the return of used aluminum cans.

1957 In its final report, the Committee on Disposal and Dispersal of Radioactive Wastes of the National Academy of Sciences recommends that the safest method for disposing of high-level radioactive wastes is underground burial in an abandoned mine or cave.

1960 The World Health Organization officially declares the end of the Third Plague.

1965 The U.S. Congress passes the first legislation dealing specifically with solid waste management in the United States, the Solid Waste Disposal Act of 1965.

1970 The U.S. Congress adopts the first legislation specifically emphasizing the role of recovery and recycling in solid waste management programs, the Resource Recovery Act of 1970.

1970 The U.S. Environmental Protection Agency is created.

1970 Hawaii resident Ruth "Pat" Webb organizes the first mass recycling program recorded in the United States. The program makes use of volunteers who pick up metal cans from roadsides, cans that are later used in the manufacture of steel bars for use in construction projects.

1971 The state of Oregon passes the first so-called bottle bill in the United States. The bill requires a deposit of five cents for each item purchased in a bottle, an amount increased to ten cents in 2017.

1971 The Nebraska Agricultural Products Industrial Utilization Committee (also known as the Gasohol Committee) is formed to find new uses for surplus grain. Its research shows that the product is an acceptable and environmentally safer substitute for conventional "leaded" gasoline. This finding is put to good use within a few years as the price of oil products from the Gulf States begins to skyrocket.

1973 The first soda bottles made out of strong, long-lasting polyethylene terephthalate (PET) become available, introducing yet another component of plastic wastes in the United States and around the world.

1975 The French Bic company introduces the first disposable razor, replacing razors with replaceable blades only. The product almost immediately becomes a huge success, one that will eventually add significant amounts to plastic trash around the world.

1976 The Toxic Substances Control Act (TSCA) of 1976 and Resource Conservation and Recovery Act (RCRA) of 1976 are adopted. TSCA is the first comprehensive legislation giving the federal government authority to regulate new and existing chemical substances. RCRA requires federal agencies to develop methods for assessing the impact of hazardous wastes on the physical, biological, and human environment.

1976 Congress adopts an extensive group of amendments to the Solid Waste Disposal Act, outlining a complete program for the treatment of solid wastes. The amendments collectively are called the Resource Conservation and Recovery Act of 1976.

The act still provides the basic framework on which the nation's solid waste management system is based.

1977 The U.S. Congress passes the Surface Mining Control and Reclamation Act of 1977, providing a system for monitoring the environmental effects of active and inactive mineral mines. One provision of the act is that the Department of the Interior develop a stream protection rule that will provide for the protection of surface water and groundwater, fish, wildlife, and other natural resources from mining wastes. The rule was updated in 1983 and 2008 and then struck down by a federal court in 2014.

1979 The EPA creates the first federal standards for sanitary landfills. Those standards deal with a range of topics including siting restrictions in floodplains, protection for endangered species, surface water and groundwater protection, disease and vector control, open burning prohibitions, explosive gas (methane) control, and fire prevention with the use of cover materials. These requirements were updated and expanded in the 1991 revisions of the RCRA.

1979 A group of homeowners in the Northwood Manor subdivision of Houston sue Southwestern Waste Management Corporation to prevent their installing a landfill in their neighborhood. The homeowners pointed to research by Dr. Robert Bullard showing that waste deposit and storage facilities in the city were disproportionately located in areas where people of color and poor people lived. The court ruled in the defendant's favor pointing out that they didn't really intend to bring about this effect.

1980 The Used Oil Recycling Act of 1980 provides regulations for collection, treatment, and disposal of used oil products in the United States.

1980 Congress passes the Comprehensive Environmental Response, Compensation, and Liability Act (CERCLA) of 1980, also known as Superfund. The act outlines a plan for identifying and cleaning up land areas that had previously been used for disposal of hazardous solid wastes.

1982 After more than twenty-five years of debate and discussion, the U.S. Congress approves the Nuclear Waste Policy Act of 1982, which outlines the steps the Department of Energy is to take in order to find one or more disposal sites for the nation's nuclear reactor waste materials.

1982 Residents of Warren County, North Carolina, supported by the United Church of Christ, stage a demonstration in opposition to the siting of a polychlorinated biphenyl (PCB) landfill near the community of Afton. More than 500 African American protestors are arrested in an unsuccessful attempt to block the construction of the landfill. This event is widely regarded as the real beginning of the modern environmental justice movement.

1986 Rhode Island becomes the first state to enact a mandatory recycling law. The law requires the recycling of all solid wastes, with fines assessed for failure to follow this regulation.

1987 Yucca Mountain, Nevada, is selected as the site for deposition and storage of the nation's high-level radioactive wastes.

1987 The Marine Plastic Pollution Research and Control Act requires the EPA and the National Oceanic and Atmospheric Administration, to study the effects of improper disposal of plastics on the environment and methods to reduce or eliminate such adverse effects.

1987 The Commission for Racial Justice of the United Church of Christ publishes a report, "Toxic Wastes and Race in the United States," showing that race, even more than income level, is the critical factor shared by communities exposed to toxic wastes.

1987 Robert D. Bullard publishes Invisible Houston, one of the first books to describe in detail environmental racism and social injustice in the black neighborhoods of Houston, Texas.

1988 The Ocean Dumping Ban Act of 1988 bans the dumping of solid wastes in the oceans and along ocean shores in the United States.

1988 The Medical Waste Tracking Act of 1988 provides mechanisms for following the location and status of medical waste materials "from cradle to grave."

1989 The Basel Convention on the Control of Transboundary Movements of Hazardous Wastes and Their Disposal is signed by fifty-three nations. It entered into force on May 5, 1992, when twenty nations had ratified the treaty.

1989 Almost a decade after being directed to do so by the Solid Waste Act Amendments of 1980, the EPA publishes a Mining Waste Exclusion list, indicating materials and processes exempt from the general conditions of the act.

1990 The Pollution Prevention Act of 1990 reiterates and strengthens the nation's commitment to the recycling and reuse of waste products in preference to all other types of waste disposal and wherever economically possible.

1990 The first national gathering on environmental justice is held at the University of Michigan in Ann Arbor. The attendees at the conference, the so-called Michigan Coalition, issue a report on "Race and the Incidence of Environmental Health." They send a letter to William Reilly, administrator of the EPA "demanding action on environmental risks in minority and low-income communities and on tribal lands." In response to the letter, Reilly appoints the Environmental Equity Workgroup.

1992 The EPA establishes an Office of Environmental Justice. One of its first activities is publication of a report on environmental justice, titled "Environmental Equity: Reducing Risks for All Communities."

1992–1993 A research team called Indigo Development, with members from Dalhousie University in Nova Scotia and Cornell University's Work and Environment Initiative, begins to explore the possibility of eco-industrial parks. Two years later, the group receives a grant from the EPA to continue and expand its research.

1994 Responding to the discovery that adequate solid waste systems were absent from all but two areas under Indian authority, Congress passes the Indian Lands Open Dump Cleanup Act to finance development and implementation of such programs.

1994 President Bill Clinton signs Executive Order 12898 on Environmental Justice ordering federal agencies to abolish and prevent policies that lead to a disproportionate distribution of environmental hazards to communities of color or low income.

1996 In one of the most successful solid waste programs in American history, Congress adopts the Mercury-Containing and Rechargeable Battery Management Act of 1996, designed to provide for labeling of all mercury- and other hazardous matter–containing batteries and elimination of the production and sale of the former. Today, more than 99 percent of lead-acid batteries are recycled.

1997 The National Recycling Coalition adopts an idea originally put into practice in the state of Texas in 1994 by declaring November 15 America Recycles Day.

2000 The Chinese government issues a directive prohibiting the import of electronic wastes for recycling. The order is largely ignored by individuals and companies involved in the activity.

2007 The Energy Independence and Security Act of 2007 is designed to increase America's independence in energy production. Among its stated objectives is an increase in the production of clean renewable fuels; an increase in the efficiency of product development, building operations, and transportation systems; and improved methods for the capture of greenhouse gases. Agricultural waste management turns out to play an important role in some of these objectives.

2007 The city of San Francisco, California, becomes the first municipality in the United States to ban the use of plastic grocery bags.

2010 The U.S. Department of Energy announces that it is abandoning consideration of Yucca Mountain as a possible site for disposal of nuclear reactor wastes.

2013 The EPA announces a policy that encourages individuals and agencies to participate in the process of granting permits for new waste disposal sites that may pose a risk for surrounding communities.

2016 The World Economic Forum and Ellen MacArthur Foundation announce the creation of the New Plastics Economy initiative. The program's goal is to create a global economy "in which plastics never become waste."

2017 China announces that it will no longer accept trash from other nations around the world for recycling.

2017 The U.S. Congress passes a new and revised version of the stream protection rule, originally issued in 1977 (*q.v.*). The action was one of the last pieces of legislation adopted by the Democratic administration of President Barack Obama. A month after being issued, the new rule was overturned by the new Republican administration of President Donald Trump.

2018 The Aquarium Conservation Partnership announces November 2018 to be "No Straw November." The purpose of the project is to encourage businesses to abandon the use of single-use plastic straws because of the threat they pose to aquatic life.

2019 One hundred eighty-six nations become signatory to the Basel Convention. Only the United States and Haiti have not yet ratified the document.

2019 The U.S. Department of Energy announces that it has decided to renew efforts to license Yucca Mountain as a disposal and storage site for nuclear reactor wastes.

2019 The EPA announces that it will be revoking an administrative rule that allows individuals and agencies to raise objections to federal decisions on the siting of hazardous waste disposal sites. This action reverses a previous rule encouraging this type of participation by stakeholders in permit decisions.

2021 The date by which Malaysia will no longer accept trash from other countries for recycling.

2025 The date by which Thailand will no longer accept trash from other countries for recycling.

A glossary of terms for the field of waste management is especially necessary because the field includes a number of widely used words that may or may not have specific technical definitions as well as terms that are used by experts in the field with precise meanings. This chapter includes examples of both. Most terms are used in the text of this book itself, while others do not occur here but may appear in other resources consulted by the reader.

aerobic composting A method of composting organic wastes using bacteria that need oxygen.

anaerobic composting A method of composting organic wastes using bacteria that do not need oxygen.

ash Noncombustible solid by-products of incineration.

baghouse A device similar to a vacuum cleaner installed in the smokestack of a factory to collect particles of hazardous waste products from a plant.

biodegradable Organic material that can be broken down by microorganisms into simpler, more stable compounds.

biogas A renewable energy source produced during the process of anaerobic digestion.

biosolids Organic materials produced by the treatment of sewage sludge. Most commonly used as fertilizers.

brownfield As defined by the U.S. Environmental Protection Agency, an "abandoned, idled, or under-used industrial and commercial facility where expansion or redevelopment is complicated by real or perceived environmental contamination."

capping The final stage of placing a covering material on a sanitary landfill.

carbon footprint The total amount of carbon (usually in the form of carbon dioxide) produced by the activities of an individual, a corporation, an activity, or some other source.

cogeneration A process by which both electricity and steam can be produced from the same incinerator or other combustion device.

commingle The process of mixing two or more solid wastes, such as paper, glass, aluminum, and plastic, in the same container. The process is also known as **single stream recycling.**

composting The process by which solid organic materials are mixed with the addition of bacteria, fungi, and other microorganisms. Degradation of the products by this method results in a rich soil-like substance.

construction and demolition wastes Waste materials produced by construction and demolition of buildings, such as bricks, concrete, drywall, metals, lumber and other wood products, and paper products. Construction and demolition wastes are characterized by their very large weight and volume.

controlled dump A waste disposal facility that makes use of some, but not all, features of a sanitary landfill. Controlled dumps often have permits that allow them to operate over limited periods of time.

dump *See* **controlled dump**, **open dump**.

electronic wastes Wastes left over from the disposal of electronic devices, such as computers, television sets, mobile phones, VCRs, copiers, and fax machines.

electrostatic precipitator A device installed on the interior walls of a smokestack that collects particles of waste gases by means of the electrostatic attraction of those particles and the device.

emission plume The shape of a body of gases or liquids that have escaped from a waste disposal site. It may occur either underground or in the atmosphere.

end destination facility Any factory, mill, or other facility at which recyclable materials are converted into new products or raw materials.

gate charge *See* **tipping fee**.

gatehouse A point of entry to a waste disposal site where fees are paid and wastes are otherwise characterized and assessed. Also called a scale house.

hazardous waste Any material that is reactive, toxic, corrosive, or otherwise dangerous to living things and the environment.

incineration The process of burning a substance, such as a waste product, for the purpose of reducing its volume and, perhaps, some of its hazardous characteristics. Heat produced by incineration is often used to produce electricity or for the heating of commercial and residential structures.

integrated solid waste management A system for dealing with waste products that takes into consideration all aspects of the process, including generation, collection, transport, treatment, recycling, incineration, and all other forms of disposal.

landfill gases Gaseous products that are produced during the decomposition of wastes in a landfill. The most common landfill gases are methane, carbon dioxide, and hydrogen sulfide.

leachate Liquid delivered to or produced by the decomposition of organic wastes within a sanitary landfill. Without methods of collecting leachate, it may leak into the groundwater, posing a threat to the environment and human health.

materials recovery facility A site at which commingled wastes are separated into individual components, such as paper, metal, and plastic, for recycling or other uses.

methane An odorless, colorless, flammable, gas with the chemical formula CH_4. It is produced during the anaerobic decomposition of organic materials.

municipal solid waste Solid waste generated as a result of residential and commercial activity, sometimes including construction and demolition wastes. The term excludes industrial, agricultural, and hazardous wastes.

National Priority List The list of Superfund sites (*q.v.*) in the United States, proposed, under repair, and deleted.

NIMBY An acronym for "not in my back yard," indicating an opposition to the siting of some undesirable facility, such as a sanitary landfill or incinerator, on a plot of land adjacent or close to a person's own property.

open dump The simplest form of waste disposal on land, consisting essentially of an open space on which wastes of any kind can be left, with no special treatment system, no method of collecting waste gases and solids, and access to anyone who wishes to use the facility.

pyrolysis Chemical decomposition of a substance by heat in the absence of oxygen.

resource recovery The extraction of usable materials and energy from wastes.

sanitary landfill A waste disposal site that contains several features designed to prevent pollution of surrounding air and water, such as gas and leachate management, proper lining, compaction of wastes, daily and final covering, access control, and record keeping.

scale house *See* **gatehouse.**

secure landfill A type of sanitary landfill with additional features designed to prevent the release of hazardous materials into the environment.

sewage sludge A semiliquid residue that settles to the bottom of canals and pipes carrying sewage or industrial wastewater. It may also form at the bottom of tanks used in treating wastewater.

single stream recycling *See* **commingle.**

site remediation The process of treating a contaminated piece of land by removing hazardous wastes that have been deposited at the site.

sludge *See* **sewage sludge.**

source reduction Any method designed to manufacture a product in such a way as to result in smaller amounts of waste.

Superfund site Any land that has been contaminated by hazardous waste and identified by the EPA as a candidate for cleanup because it poses a risk to human health or the environment.

sustainability As defined by the United Nations World Commission on Environment and Development, sustainability means "development that meets the needs of the present without compromising the ability of future generations to meet their own needs."

tipping fee The charge assessed for unloading waste materials at a dump, transfer station, landfill, incinerator, or other disposal site.

transfer station A facility at which solid wastes are temporarily deposited for later transport to more distant, more permanent disposal sites.

universal wastes A special and limited category of wastes that includes batteries, lamps, mercury-containing equipment, and pesticides.

waste picker A person who picks out recyclables from mixed waste in dumps, landfills, or other waste disposal sites.

waste stream A term used to describe the total flow of waste materials from households, office buildings, and other structures to their final disposal point.

waste-to-energy plant A facility at which wastes are burned to produce heat or electrical energy for use at some other location.

white goods Discarded appliances, such as washing machines, dryers, stoves, refrigerators, and fencing.

yard trimmings Grass, leaves, brush, tree branches and stumps, and other plant material from residential, institutional, and commercial sources.

zero waste Any system designed to reduce, if not eliminate, the amount of solid wastes produced, using modifications of current procedures for generation of a material or product, reusing and recycling it, or reprocessing it to produce useful products.

Page numbers followed by *t* indicate tables.

About the Author

David E. Newton holds an associate's degree in science from Grand Rapids (Michigan) Junior College, a BA in chemistry (with high distinction), an MA in education from the University of Michigan, and an EdD in science education from Harvard University. He is the author of more than 400 textbooks, encyclopedias, resource books, research manuals, laboratory manuals, trade books, and other educational materials. He taught mathematics, chemistry, and physical science in Grand Rapids, Michigan, for thirteen years; was professor of chemistry and physics at Salem State College in Massachusetts for fifteen years; and was adjunct professor in the College of Professional Studies at the University of San Francisco for ten years.

The author's previous books for ABC-CLIO include *Global Warming* (1993), *Gay and Lesbian Rights* (1994, 2009), *The Ozone Dilemma* (1995), *Violence and the Media* (1996), *Environmental Justice* (1996, 2009), *Encyclopedia of Cryptology* (1997), *Social Issues in Science and Technology: An Encyclopedia* (1999), *DNA Technology* (2009, 2016), *Sexual Health* (2010), *The Animal Experimentation Debate* (2013), *Marijuana* (2013, 2017), *World Energy Crisis* (2013), *Steroids and Doping in Sports* (2014, 2018), *GMO Food* (2014), *Science and Political Controversy* (2014), *Wind Energy* (2015), *Fracking* (2015), *Solar Energy* (2015), *Youth Substance Abuse* (2016), *Global Water Crisis* (2016), *Same-Sex Marriage* (2011, 2016), *Sex and Gender* (2017), *STDs in the United States* (2018), *Natural Disasters* (2019), *Vegetarianism and Veganism* (2019), *Perspectives* (editor, 2019), *Eating Disorders*

(2019), and *Gender Inequality* (2019). His other recent books include *Physics: Oryx Frontiers of Science Series* (2000); *Sick!* (4 volumes, 2000); *Science, Technology, and Society: The Impact of Science in the 19th Century* (2 volumes, 2001); *Encyclopedia of Fire* (2002); *Molecular Nanotechnology: Oryx Frontiers of Science Series* (2002); *Encyclopedia of Water* (2003); *Encyclopedia of Air* (2004); *The New Chemistry* (6 volumes, 2007); *Nuclear Power* (2005); *Stem Cell Research* (2006); *Latinos in the Sciences, Math, and Professions* (2007); and *DNA Evidence and Forensic Science* (2008). He has also been an updating and consulting editor on a number of books and reference works, including *Chemical Compounds* (2005); *Chemical Elements* (2006); *Encyclopedia of Endangered Species* (2006); *World of Mathematics* (2006); *World of Chemistry* (2006); *World of Health* (2006); *UXL Encyclopedia of Science* (2007); *Alternative Medicine* (2008); *Grzimek's Animal Life Encyclopedia* (2009); *Community Health* (2009); *Genetic Medicine* (2009); *The Gale Encyclopedia of Medicine* (2010–2011); *The Gale Encyclopedia of Alternative Medicine* (2013); *Discoveries in Modern Science: Exploration, Invention, and Technology* (2013–2014); and *Science in Context* (2013–2014).